The Constituents of Medicinal Plants

3rd Edition

The Constituents of Medicinal Plants

3rd Edition

Andrew Pengelly, PhD

CABI is a trading name of CAB International

CABI
Nosworthy Way
Wallingford
Oxfordshire OX10 8DE
UK

Tel: +44 (0)1491 832111
Fax: +44 (0)1491 833508
E-mail: info@cabi.org
Website: www.cabi.org

CABI
WeWork
One Lincoln St
24th Floor
Boston, MA 02111
USA

Tel: +1 (617)682-9015
E-mail: cabi-nao@cabi.org

A catalogue record for this book is available from the British Library, London, UK.

Library of Congress Cataloging-in-Publication Data

Names: Pengelly, Andrew, author.
Title: The constituents of medicinal plants / Andrew Pengelly.
Description: Third edition. | Boston, MA, USA : CAB International, [2021] |
    Includes bibliographical references and index. | Summary: "Pengelly's book is a
    classic in the literature of herbal medicine. It is an easy to understand introduction
    to the chemistry of medicinal plants. This new edition contains six new chapters
    incorporating topics of contemporary interest, including cannabinoids, mushroom
    polysaccharides and mycochemistry, and toxicology of phytochemicals"-- Provided
    by publisher.
Identifiers: LCCN 2020046262 (print) | LCCN 2020046263 (ebook) | ISBN
    9781789243079 (paperback) | ISBN 9781789243086 (ebook) | ISBN
    9781789243093 (epub)
Subjects: LCSH: Herbs--Therapeutic use. | Medicinal plants--Analysis.
Classification: LCC RM666.H33 P465 2021 (print) | LCC RM666.H33 (ebook) |
    DDC 615.3/21--dc23
LC record available at https://lccn.loc.gov/2020046262
LC ebook record available at https://lccn.loc.gov/2020046263

References to Internet websites (URLs) were accurate at the time of writing.

ISBN-13: 9781789243079 (paperback)
        9781789243086 (ePDF)
        9781789243093 (ePub)

DOI: 10.1079/9781789243079.0000

Commissioning Editor: Rebecca Stubbs
Editorial Assistant: Lauren Davies
Production Editor: James Bishop

Typeset by SPi, Pondicherry, India
Printed and bound by CPI Group (UK) Ltd, Croydon, CR0 4YY

Dedication

I dedicate this book to my late friends and mentors Dr Art Tucker and Dr Jim Duke, two of the giants of phytochemical and medicinal plant research.

Contents

Foreword

The scientific understanding of therapeutic plants is rapidly growing and changing, meaning it is vital that students and therapists in the field keep abreast of these developments for the benefit of their current and future patients. Fundamental to this is the need to have accurate and up-to-date text books. In this context, the third edition of *The Constituents of Medicinal Plants* by Andrew Pengelly has addressed a crucial need and is a welcome addition to my library. Our recent understanding of the therapeutic uses of plants has revealed several significant issues that have the potential to impact on the quality, safety and efficacy of herbal products. It is essential that all practitioners and students of herbal medicine, whatever their philosophical leanings, have the tools to understand and manage these issues. An effective understanding of the science of modern herbal practice begins with a sound knowledge of the phytochemistry and related therapeutics of medicinal plants.

This new edition from Andrew is a comprehensive introduction to the main phytochemical classes found in plants that are understood to confer their therapeutic activity. A major advantage is that the book assumes only a basic understanding of chemistry, making it an ideal primer for students and practitioners alike. However, it quickly brings the reader into more sophisticated territory, without falling into the trap of being overly reductionist or technical. Key features are the many chemical diagrams and the wide-ranging discussion of pharmacological activities and safety considerations.

The new third edition has been thoroughly revised to incorporate the latest research. It contains a new chapter on resins and cannabinoids, and additional content on essential oil chemotypes, the important area of mushroom polysaccharides, together with a discussion of phytochemical synergy and updates on the toxicology of plant constituents.

The author is a well-known and respected authority on medicinal plants with decades of experience in the field. Through his teaching in Australia and overseas, Andrew has helped to pioneer the scientific approach to herbal practice, without losing sight of its traditional underpinnings.

Kerry Bone
Director of R&D, MediHerb
Adjunct Professor, New York Chiropractic College, School of Applied Nutrition

Preface to the Third Edition

As noted in the preface to the first (1996) edition of *The Constituents of Medicinal Plants*, this book was originally designed as a formal set of teaching notes for my students in herbalism and naturopathy, many of whom had little previous training in chemistry or the biological sciences. Since then the book has become something of a standard reference text at universities and colleges that teach some form of plant-based medicine, as well as within the research community. The second (2004) edition is now over 16 years old, however, I still receive almost daily notifications from the academic networking site ResearchGate, of citations to the text in peer-reviewed publications – over 150 such citations are listed on Google Scholar. Therefore, it was a pleasant surprise when I was invited by CABI Publishing, the UK-based distributor for the second edition, to write this third edition.

In the last 20 years the state of published research in the field of phytochemistry has expanded exponentially, the literature now being much more readily available, due to the increased levels of open access content appearing in peer-reviewed journals. Comprehensive literature reviews for many of the main phytochemical groups and the plants from which they come are now available. These days the problem confronting academics and researchers is more likely to be a case of information overload than lack of information. However, the information is often spread across a broad range of journals and textbooks each with their own specialty. These include pharmacy, pharmacology, chemistry, plant biology, molecular biology, herbal medicine, integrative medicine, food science, natural products research and others. Hence, I realized that my role would be to review as much of the published literature as I could, then to condense the information into a more reader-friendly format. I leave it to the readers and reviewers to decide to what extent I have succeeded in this mission.

The format of the third edition has changed little from previous editions, as before the names of key phytochemicals used in herbal medicine are set in bold, usually on first mention in each chapter. One chapter has been added, that of 'Resins and Cannabinoids' which were previously lumped in with essential oils, but which clearly have unique characteristics not shared with essential oils. Cannabinoids, in particular, have little in common with essential oils other than their lipophilic (fat soluble) properties. The new chapter also reflects on the growing interest of cannabinoids within the research community, and they are something of a posterchild for phytochemistry, originating as they do from multiple biosynthetic pathways.

Other chapters have been considerably expanded. For example, the polysaccharide chapter now includes more comprehensive sections on seaweeds and fungi than in previous editions, emphasizing in particular their immunomodulatory effects. Seaweeds and mushrooms are excellent examples of dietary ingredients that have been consumed by humans for centuries, with a good track record for safety. They have both demonstrated potent effects for prevention and treatment of cancer and other chronic illnesses. These are sometimes labelled as 'functional foods' and marketed as dietary supplements, however, the benefits of seaweeds and mushrooms may still be obtained from natural dietary sources. Similarly, the tannins chapter is expanded to include more polyphenol-rich dietary ingredients such as pomegranate, grapes and Kakadu plum. Modern science reveals that these foods of the ancients contain extremely high levels of polyphenols, which have been shown to impart powerful antioxidant and anti-inflammatory actions, among others.

While the focus of a text such as this is primarily concerned with the molecular structures of the 'active constituents' in medicinal plants and their therapeutic effects, it is not my belief that this reductionist approach is the only way to explain or rationalize plant-derived medicines. Much can also be learned from energetics, the philosophy that underpins the origins of both Eastern and Western systems of medical practice, not to mention some modern approaches to the biological sciences that emphasize complexity, including the concept of synergism and the emerging field of network pharmacology. Liu, in the *Journal of Nutrition*, stated 'The additive and synergistic effects of phytochemicals in fruit and vegetables are responsible for their potent antioxidant and anti-cancer activities' (Liu, 2004). With some exceptions I do believe that the biological effects of medicinal plants lie in the interactions between their various phytochemicals and their molecular targets, as distinct from the silver-bullet approach of a single chemical targeting a single receptor.

I would like to acknowledge the support of my publishers at CABI Publishing, in particular Rebecca Stubbs, for approaching me to write this edition in the first place, and her graciousness and patience when I repeatedly missed submission deadlines. For she who hates clichés, I couldn't have completed this work without the support of my life-partner Kathleen Bennett. Kat's background in professional writing and editing helped rescue some of my clumsy phraseology, and she ensured that every reference and citation was diligently checked for accuracy and appropriateness, before the manuscript was put into the publisher's hands. So, a heart-felt 'thank you' goes out to both Rebecca and Kathleen. I am forever indebted to my most influential teacher, Denis Stewart founder of Southern Cross Herbal School, who instilled the notion in me that herbal medicine was part medicine, part science and part philosophy. Finally, as always, it is the readers – students, herbalists, academics and researchers from all over the world – who have given me such positive feedback over the years, and for whom this new edition has been written.

Reference

Liu, R.H. (2004) Potential synergy of phytochemicals in cancer prevention: mechanisms of action. *Journal of Nutrition* 134, 3479S–3485S.

Introduction to Phytochemistry 1

Introduction

The field of medicine has long been divided between so-called 'rationalist' and 'vitalistic' principles. While the rationalist/scientific model has held sway (at least in the Western nations) for the last couple of centuries, vitalistic concepts of health and healing have made a comeback in recent decades. A vast array of natural healing modalities – both ancient and new – has emerged, and some are even challenging orthodox medicine for part of the middle ground. Alternative medicine has become 'complementary and alternative medicine' (CAM) or 'complementary medicine' (CM), while more recently 'integrative medicine' is seen as a blend of orthodox medicine and CAM. However, the question is often asked: 'Is there any scientific evidence that proves any of these therapies work?'

Of all the various complementary therapies, perhaps medical herbalism can be made to fit the orthodox model most easily. Given that many of the pharmaceutical drugs in use are derived from plants directly or indirectly, it is obvious that at least some plants contain compounds with pharmacological activity that can be harnessed as medicinal agents. While few would disagree with that proposition, there are many who persist in referring to herbal medicines (along with other 'alternative remedies') as unproven and therefore of little or no clinical value. Increasingly, the public – and particularly the medical establishment – are demanding herbalists and other complementary therapists provide scientific evidence for the efficacy and safety of their practices. While this is an admirable objective, it cannot be achieved overnight given the complexities of the herbs themselves, the variety of formulas and prescribing methods available, and the difficulties in adapting biomedical models to the herbal practice.

Indeed, there are many inside the medical establishment who question the validity of double-blind controlled trials and 'evidence-based medicine' in general (e.g. Black, 1996; Vincent and Furnham, 1999). In a formal evaluation procedure, the quality of randomized controlled trials of interventions using complementary medicines was found to be more or less the same as those

DOI: 10.1079/9781789243079.0001

1

using conventional biomedicine – although the overall quality of evidence in both cases was generally regarded as poor (Bloom *et al.*, 2000). This assessment supports the point made by Black that 'the difference in the standards of evidence for orthodox and complementary therapies may not be as great as generally assumed' (Black, 1996). More recently large umbrella bodies have been set up in Europe (the CAMbrella project), the USA (the National Center for Complementary and Alternative Medicine (NCCAM) Strategic Plan) and the World Health Organization with objectives that include further integration of CAM and conventional healthcare, prioritization of research, to improve the capacity and rigour of research and to disseminate objective, evidence-based information on CAM treatments (Fischer *et al.*, 2014; Zhang *et al.*, 2019).

Phytochemical basis of herbal medicines

Since herbal medicines are products of the biological world, their properties and characteristics can be studied using the accumulated skills and knowledge embedded in the natural sciences – especially botany, chemistry and molecular biology. Through an understanding of simple principles of chemistry, we see there is a similarity in the molecules that make up plants and humans, while foods and medicines derived from plants provide a chemical continuum between these two kingdoms. The more we comprehend these natural processes, the easier it is for us to intervene using biological agents (in this case herbs) to alleviate diseased states in our fellow humans.

To a pharmacist, a plant's constituents may be regarded as an unholy mixture of mainly unwanted chemicals, to be refined with the objective of identifying and isolating an 'active principle'. Herbalists on the other hand aim at a holistic approach – one that values the sum or totality of a plant's constituents – even those considered by the pharmacist to be worthless. In order to study the activity of a given herb, it is often necessary to purify it or isolate a specific compound – an example of the reductionist approach that characterizes the biomedical model.

While many of the studies referred to in this book are a product of such reductionist research, the results or findings should not be devalued in principle. Isolation of and experimentation with single constituents provides information that can be adapted to a more holistic understanding of a herb's action. Knowledge of individual constituents is also essential for developing quality assurance methods, extraction procedures, understanding of pharmacological activity and pharmacokinetics and – most importantly – understanding of potential toxicology and interactions with pharmaceutical drugs. It is not merely a necessary step in the isolation and synthesis of plant-derived drugs.

Understanding organic chemistry

It does not require a science degree to gain an understanding of the fundamental chemical structures found in medicinal herbs, but some knowledge of organic chemistry is desirable. Hence reference to any good introductory text on organic chemistry or biochemistry will help those who haven't done an elementary course at tertiary level, and there are many excellent primers for organic chemistry on the World Wide Web.

I am indebted to some of the great scientists and herbalists who have inspired me with their knowledge of the subject, making the job of learning phytochemistry much easier for the non-chemists – teacher, student and practitioner alike. These include Michael Wink and Ben-Erik van Wyk, Hilderbert Wagner, Paul Dewick, N.G. Bissett, James Duke, Jean Bruneton, Alexander Panossian and Arthur O. Tucker. I highly recommend the publications of these authors many of which are listed in the suggested reading.

In this chapter we review some of the basic chemical principles and terminology that are used throughout the book, provide an introduction to the biosynthetic processes through which plants manufacture their chemicals, and explore some of the recent investigations into synergism between medicinal plants and their constituents.

Biosynthesis of organic compounds

Photosynthesis

Photosynthesis is a process by which the leaves of plants manufacture carbohydrates and oxygen, using carbon dioxide from the air and water absorbed from the roots. The following equation should be familiar to anyone who studied biology at high school.

$$6CO_2 + 6H_2O = C_6H_{12}O_6 + 6O_2$$

This reaction is only possible under the influence of sunlight and in the presence of specialized organelles, found in plant cells, known as chloroplasts, which contain the light-trapping pigment chlorophyll.

Biosynthetic pathways

Virtually all chemical compounds found in plants derive from a few well-studied metabolic pathways. The 'pathways' begin with chemical products of photosynthesis and glycolysis (glucose metabolism) – simple starting molecules (precursors) such as pyruvic acid, acetyl coenzyme A and organic acids. A series of intermediate compounds are formed which are quickly reduced – with the assistance of specific enzymes – into other, often unstable intermediate compounds, until finally a complex, stable macromolecule is formed. Metabolic pathways involve a series of enzymes specific for each compound.

Primary and secondary metabolites

The biosynthetic pathways are universal in plants and are responsible for the occurrence of both primary metabolites (carbohydrates, proteins, etc.) and secondary metabolites (phenols, alkaloids, etc.). Secondary compounds were once regarded as simple waste products of a plant's metabolism. However, this argument is weakened by the existence of specialist enzymes, strict genetic controls and the high metabolic requirements of these compounds (Waterman and Mole, 1994). Most scientists accept that many of these compounds serve primarily to repel grazing animals or destructive pathogens (Cronquist, 1988; Kabera *et al.*, 2014), while additional functions most likely include acting as signalling compounds to attract pollinators and for protection from ultraviolet (UV) radiation (Wink, 2015).

Biosynthetic reactions are energy consuming, fuelled by the energy released by glycolysis of carbohydrates and through the citric acid cycle. Oxidation of glucose, fatty acids and amino acids results in formation of ATP (adenosine triphosphate), a high-energy molecule formed by catabolism (enzymic breakdown) of primary compounds. ATP is recycled to fuel anabolic (enzymic synthesis) reactions involving intermediate molecules on the pathways.

Whereas catabolism involves oxidation of starting molecules, biosynthesis or anabolism involves reduction reactions, hence the need for a reducing agent or hydrogen donor, which is usually NADP (nicotinamide adenine dinucleotide phosphate). These catalysts are known as coenzymes and the most widely occurring is coenzyme A (CoA), made up of ADP (adenosine diphosphate) and pantetheine phosphate.

The main biosynthetic pathways are:

- pentose – fatty acids, aromatic amino acids, polysaccharides;
- shikimic acid – phenols, tannins, aromatic alkaloids;
- acetate–malonate – phenols, alkaloids; and
- mevalonic acid – terpenes, steroids, alkaloids.

The structures of organic compounds

Of elements and atoms

An element is a substance that cannot be divided further by chemical methods – it is the basic substance upon which chemical compounds are built. The Periodic Table classes all known elements in a systematic manner based on the increasing number of electrons and protons (which are equal), starting with hydrogen (number 1: since it has one electron and one proton).

Atoms are the smallest particle within elements. They are made up of protons and neutrons (in the nucleus) and electrons (in orbits around the nucleus). Each orbit represents an energy level and these give the atom stability.

Electrons in the outer orbit, or valence shell, control how the atom bonds. When atoms are linked together by chemical bonds, they form molecules.

To achieve chemical stability, an atom must fill its outer electron shell, and it does this by losing, gaining or sharing electrons. These are known as valence electrons and the valence is specific for each element.

Chemical bonds

A bond is a pair of electrons shared by the two atoms it holds together. There are many types of chemical bonds including hydrogen, ionic and covalent bonds. In organic chemistry (based on the element carbon) we deal mainly with covalent bonds, which may occur as single, double or triple bonds.

Covalent bonds have a shared pair of electrons between two atoms – they neither gain nor lose electrons, as ionic bonds do. They occur in elements towards the centre of the Periodic Table, the most significant element from the perspective of phytochemistry being carbon. Covalent bonds are stronger than hydrogen or ionic bonds and covalent compounds don't form solutions with water. Covalent bonds may be polar or non-polar depending on the relationship between the electric charges emitted by the respective atoms.

The bonding properties of elements are related to their valence, that is, the number of electrons they need to fill their outer shells. The most abundant elements found in living organisms are:

Hydrogen   (H)
Oxygen   (O)
Nitrogen   (N)
Carbon   (C)

The bonding properties, or valence bonds, are 1, 2, 3, 4, respectively, hence the HONC rule (Perrine, 1996):

H forms 1 bond
O forms 2 bonds
N forms 3 bonds
C forms 4 bonds

From the HONC rule we learn that carbon must always be linked to other atoms through four bonds. For example, the formula for methane is CH_4. We can draw it in a way that represents the bonding arrangement:

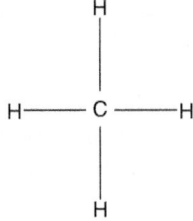

Acyclic, cyclic and heterocyclic compounds

The atoms of organic compounds are arranged as either open chains (acyclic or aliphatic) or closed ring systems (cyclic). In molecular structure diagrams, each corner or kink in the ring (or chain) indicates a CH_2 group, although these are usually abbreviated to C or omitted. Each line represents a bond.

Unsaturated ring systems are those in which the carbons are linked by double or triple bonds, while saturated rings do not contain any double bonds. In the diagram below, the cyclohexane ring is a saturated ring with each carbon labelled. The benzene ring, the central structure of thousands of organic compounds, is an unsaturated six-carbon ring, generally illustrated as a hexagon containing three double lines for the conjugated (alternating) double bonds. The labels are omitted in this example. Compounds containing one or more benzene rings are known as aromatic compounds.

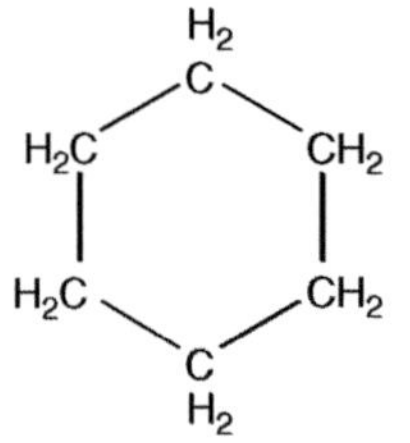

cyclohexane ring    benzene ring

Once you have looked at these structures often enough, the labelling of atoms is unnecessary – since only one arrangement of atoms is possible for each bonding configuration according to the HONC rule. See if you can count the number of bonds held by each carbon atom in the cyclohexane ring above – there must be four.

Ring systems in which the rings are composed entirely of hydrocarbons (CH_2) are called homocyclic – the benzene ring is a good example. Ring systems containing one or more additional elements are called heterocyclic. In phytochemicals, the most common additional elements are nitrogen, oxygen and sulfur (often spelt sulphur). Over 4000 heterocyclic systems are known from plant and animal sources. They sometimes occur fused to a benzene ring or to another heterocyclic ring, to give bicyclic systems. Some of these heterocyclic rings resist opening and remain intact throughout vigorous reactions, as does the benzene ring.

About half of the over 12 million registered chemical compounds contain heterocyclic ring systems (Kaur et al., 2013). Tannins are good examples, in the structure of ellagic acid in the figure below, four heterocyclic rings are fused together.

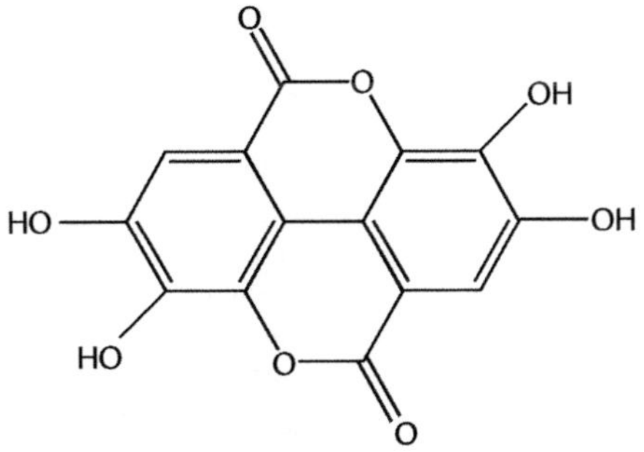

ellagic acid

Some important parent heterocyclic compounds are shown below:

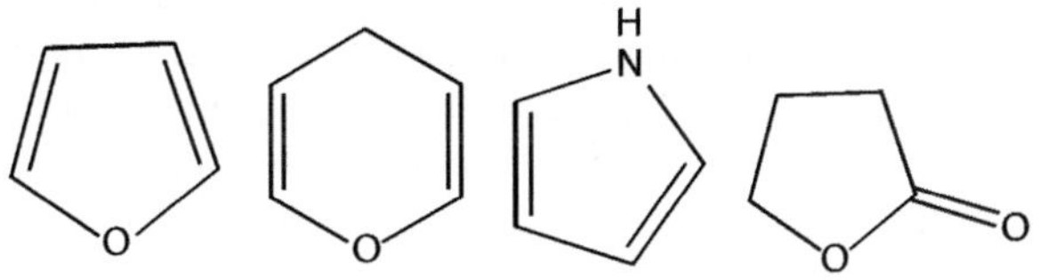

furan   pyran   pyrrole   lactone

Functional groups

Another feature of organic compounds is the presence of functional groups. These are groups of atoms attached to carbon chains or rings which are often involved in chemical reactions. The different classes of functional groups are distinguished by the number of hydrogen atoms they replace or substitute. Common ones include alcohol, ketone, aldehyde, amine and carboxylic acid (see the diagram below).

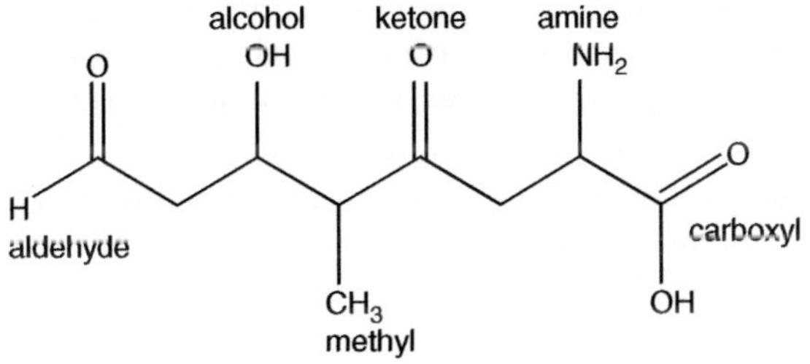

imaginary hydrocarbon chain with functional group attachments

That a compound includes a particular functional group is indicated by a prefix or suffix in the name of the compound (Table 1.1).

Table 1.1. Prefixes and suffixes for functional groups in organic chemistry

Functional group	Prefix	Suffix
Carboxylic acids	None	-oic acid
Aldehydes	None	-al
Ketones	None	-one
Alcohols	hydroxy-	-ol
Amines	amino-	-amine
Amides	None	-amide
Ethers	alkoxy-	-ether
Thiols	allyl-	-thiol
Alkyl	None	-yl

Functional groups are usually labelled in the shorthand manner shown above – even though the hydrocarbon chain (or ring) remains unlabelled. Functional groups exert a significant influence on the behaviour of molecules, especially small molecules such as those found in essential oils. Some of the more reactionary functional groups, including many alkyls and thiols, have the additional property of forming covalent bonds with proteins and peptides such as enzymes and receptors, thereby inducing physiological changes.

The systematic naming of molecules is one of chemistry's great mysteries, certainly to the uninitiated. As is the case for botanical names, there are a set of rules set by an international organization, which is reviewed and updated in peer-reviewed journals. International Union of Pure and Applied Chemistry (IUPAC) rules may be found in textbooks of organic chemistry, with updates in journals published by De Gruyter. For example, the nomenclature of flavonoids was recently published as open access in the journal *Pure and Applied Chemistry* (Rauter *et al.*, 2018). Fortunately, commonly referred-to molecules are usually given common names. In the example below, the term 'flavonol' is used as a class name for compounds with a 3-hydroxy-2-phenyl-4H-1-benzopyran-4-one skeleton.

In general, the name of a compound is written with the substituents or functional groups in alphabetical order, followed by the base name (derived from the number of carbons in the parent chain). Commas are used between numbers and dashes are used between letters and numbers (Sarda *et al.*, 2017). Despite these rules, there are often two to three different chemical names that can apply to the same molecule. Fortunately for the reader, this textbook uses the common names of molecules whenever possible, however, as with botanical names, the systematic names as agreed to by the IUPAC can provide useful information about a molecule.

Isomerism

The mystery surrounding organic chemical structures is partly due to the three-dimensional shapes of these molecules: carbons tend to form tetrahedrons rather than planar (two-dimensional) structures. This allows for two or more positions of atoms on the same basic molecule. There are several types of isomers.

Constitutional isomers

Constitutional isomers are compounds with the same molecular formula but a different connectivity of bonded atoms.

In terpenes, the positioning of a double bond is indicated by prefixing the name of the compound with alpha (α), beta (β), gamma (γ) or delta (δ) – as in the example of **terpinene** (see illustration). Another example comes from the triterpenoids **α- and β-amyrin** (see illustration). These are parent compounds for a different set of triterpenoid saponins, for example α-amyrin is the parent molecule of **asiatoside** and **asiatic acid** from *Centella asiatica* (Apiaceae), **ursolic acid** from *Salvia rosmarinus* (Lamiacaeae) and faradiol monoester from *Calendula officinalis* (Asteraceae), while the β-amyrin is the parent of glycyrrhizin from *Glycyrrhiza glabra* (Fabaceae), saikosaponin from *Bupleurum falcatum* (Apiaceae), α-hederin from *Hedera helix* (Araliaceae) and soyasaponin from *Glycine max* (Fabaceae). This is a clear example in which compounds with the same molecular formula, but different connections, can have different biological properties.

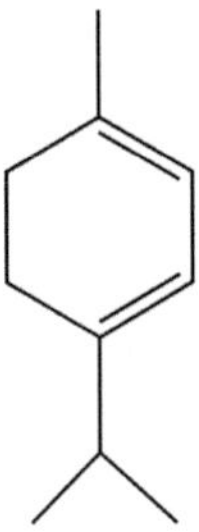

α-terpinene 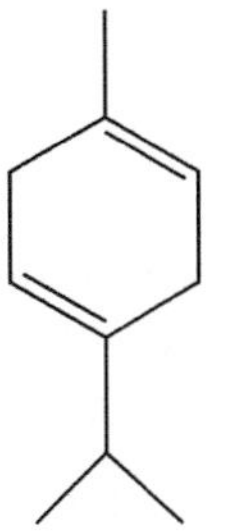   γ-terpinene

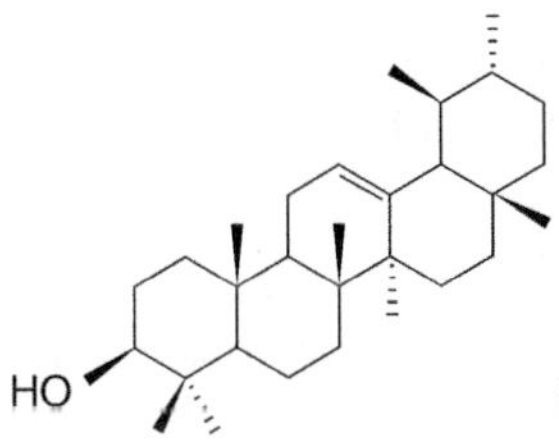

α-amyrin 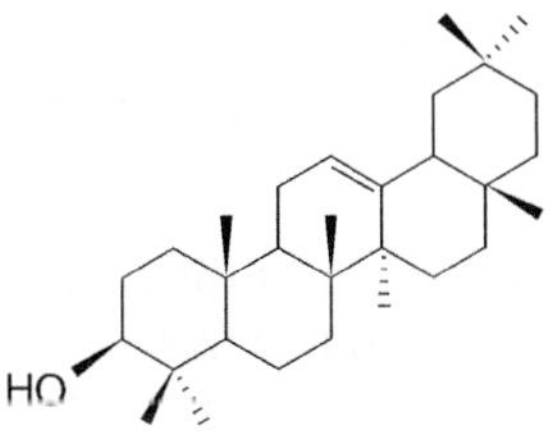   β-amyrin

Positional isomers

Positional isomers differ in the position of their functional group. They may be compounds whose side chains are attached at different locations around the carbon ring. For example, the phenol coumaric acid may contain an hydroxyl (OH) group at any of three locations, known as *ortho* (*o*-coumaric acid), *meta* (*m*-coumaric acid) or *para* (*p*-coumaric acid). **Thymol** and **carvacrol** are positional isomers due to the different positions of the hydroxyl group on the monoterpene skeleton. The two compounds have similar but not identical actions.

In modern chemistry the positional isomer delegations (*ortho, meta* and *para*) are becoming obsolete, since the positions can be indicated by IUPAC numbering system. Hence in the case of **terpinen-4-ol**, the major constituent of tea tree oil, the number 4 designates the position of attachment of the hydroxyl group to the ring, and the term '*para*' is not required.

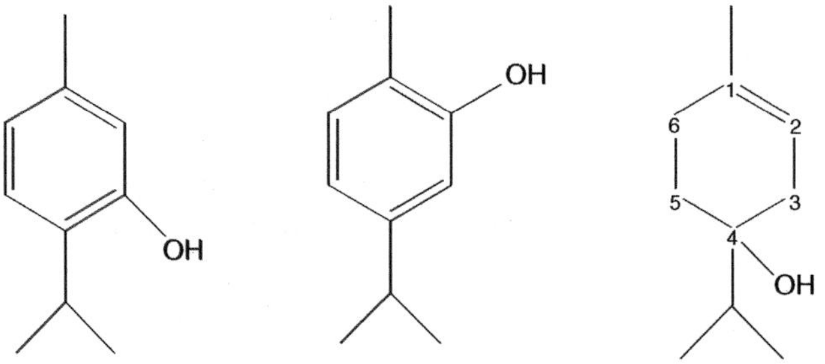

thymol   carvacrol   terpinene-4-ol

Stereoisomers

Stereoisomers have the same bonds or connectivity, but a different three-dimensional orientation of atoms. Geometric (*cis–trans*) isomers differ in the placement of functional groups on one side or other of the double bond:

1. *Cis* designates the stereoisomer with like groups on the same side of the double bond.
2. *Trans* designates the stereoisomer with like groups on opposite sides of the double bond.
3. *Cis–trans* isomerism is responsible for significant differences in the properties and odours of many essential oils containing identical chemical constituents.

In modern chemistry texts the *cis/trans* nomenclature has been replaced by a notational system known as *E–Z* where Z corresponds to *cis* and *E* to *trans*, although there are exceptions to this correlation. In this system, when the higher atomic number atoms are on opposite sides of the double bond the configuration is *E*.

The organic acids maleic acid and fumaric acid are *cis–trans* or *E–Z* isomers (see diagram below), while cinnamaldehyde – responsible for the odour of cinnamon – occurs in the *trans* or *E* form only.

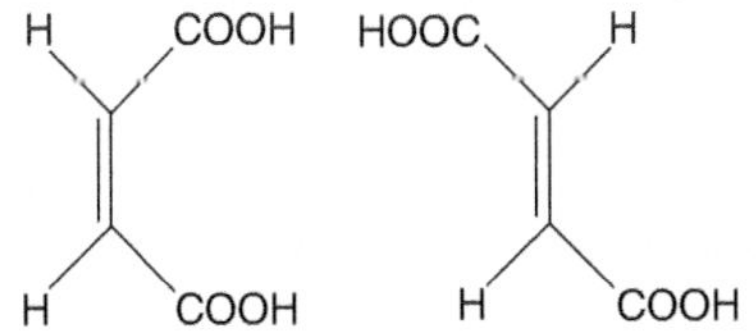

maleic acid (*cis*) (*Z*)   fumaric acid (*trans*) (*E*)

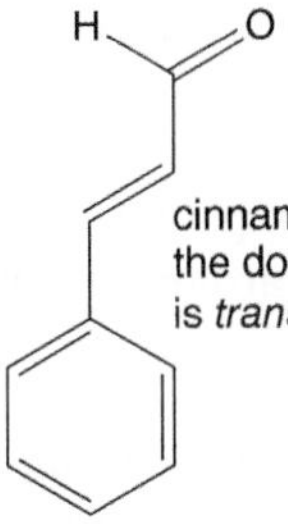

cinnamaldehyde:
the double-bonded carbon
is *trans* or *E* configuration

Enantiomers

These are non-superimposable mirror images known as chiral[1] molecules.

In tetrahedrons, chirality of a molecule occurs when the carbon atom is bonded to four different atoms or functional groups Their different configuration is designated as (*R*)- or (*S*)- by the Cahn Ingold Prelog (CIP) system for naming chiral centres.

Wright (2004) provides examples of chiral pharmaceutical drugs that are only effective in one enantiomeric form. The anti-inflammatory drug ibuprofen is only active as (*S*)-ibuprofen, while the notorious teratogenic drug thalidomide is toxic only as (*R*)-thalidomide, the (*S*)- form does not harm fetuses.

To distinguish between enantiomers of monosaccharides and amino acids, an older convention using L and D is still in use. Isomeric molecules are represented as 'Fischer projections', two-dimensional depictions with a carbon chain backbone and functional groups set out to the left and right. The position of specific functional groups in relation to a chiral carbon dictates whether the molecule is labelled L or D. All essential amino acids have the key functional amino group occurring on the left side of the carbon chain, the L-configuration (e.g. L-lysine, L-tyrosine). D-amino acids are commonly found in bacteria.

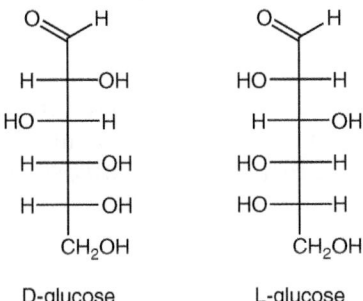

D-glucose   L-glucose

Fischer projections demonstrate stereoisomerism in glucose

Enantiomers are also optical isomers – molecules that rotate plane-polarized light by identical magnitudes but in different directions:

1. Dextrorotary: *d* or (+)- rotates light clockwise (to the right).
2. Laevorotary: *l* or (–)- rotates light anticlockwise (to the left).
3. Racemic mixture: *dl* or (±)- – the mixture includes an equal amount of enantiomers.

Following IUPAC nomenclature, *l* and *d* notation has been replaced by (+)- and (−)- symbols.

Essential oil monoterpenes offer some good examples of the influence of optical isomerism. For example **(+)-carvone** and **(−)-carvone** are responsible for the familiar but quite different aromas of caraway seed and spearmint, respectively.

Diastereomers
Diastereomers are stereoisomers that do not have a mirror image relationship. These molecules have more than one chiral centre. Tartaric acid, maleic and fumaric acids are examples of diastereomers.

Organic acids

Organic acids are of such widespread occurrence that they are not strictly secondary metabolites; in fact many occur during the citric acid or Krebs cycle. They are water-soluble colourless liquids with characteristic sharp tastes that influence the flavour of most of the fruit we eat.

Monobasic acids contain a single carboxyl group (COOH). They include the fatty acids as well as **isovaleric acid**, a sedative principle found in *Valeriana officinalis* (Caprifoliaceae) and *Humulus lupulus* (Cannabaceae). One of the most important in this group is **acetic acid**, the main constituent in vinegar. Acetic acid is a precursor for the biosynthesis of lipids, as well as some essential oils and alkaloids.

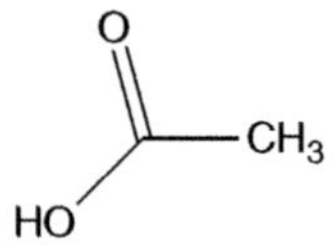

acetic acid

Polybasic acids contain two or more carboxyl groups and generally have a slight laxative effect. They are **oxalic acid** – a common ingredient of vegetables, **succinic acid** found in the root of dang gui (*Angelica sinensis*, Apiaceae), and **fumaric acid** found in fumitory (*Fumaria officinalis*, Papaveraceae).

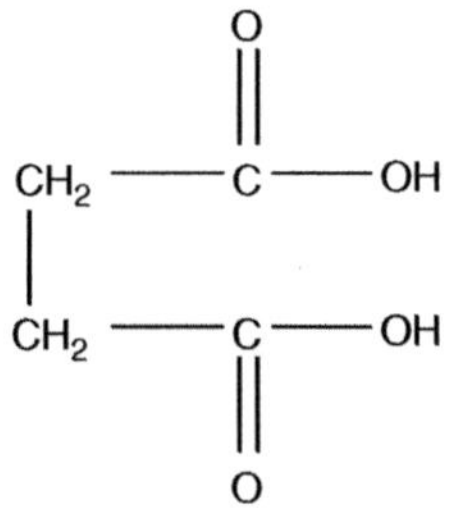

succinic acid

Hydroxyl acids include a hydroxyl group (OH) with a pair of carboxyl groups. **Citric acid** and **tartaric acid** are the most common examples. **Lactic acid** is an exception in that it has only one carboxyl group. Lactic acid is enantiomeric – racemic forms occur in soured milk products.

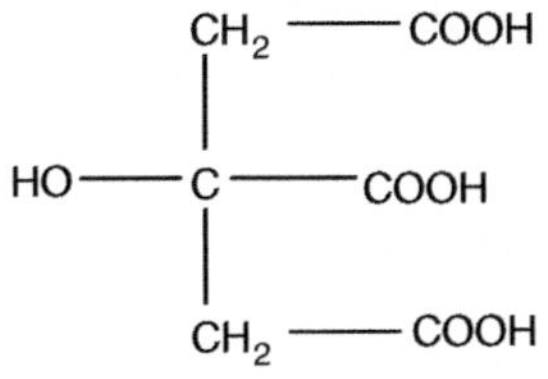

$$CH_2 \text{——} COOH$$
$$HO \text{——} C \text{——} COOH$$
$$CH_2 \text{——} COOH$$

citric acid

Aromatic acids such as benzoic acid are sometimes classed with the organic acids; however, they are products of the shikimic acid pathway and are discussed further in Chapter 2, this volume.

Organic acids have antibacterial effects, most notably in the urinary tract. Cranberry (*Vaccinium macrocarpon*, Ericaceae) is rich in organic acids, and these are beneficial in treating bladder infections. In one study cranberry juice significantly reduced the number of colony-forming units of *Escherichia coli* in the bladder of experimental mice, this action was attributed to the combination of **malic**, citric, **quinic** and **shikimic acids** (Jensen *et al.*, 2017).

Synergism

While this book by necessity deals, in the main, with the properties of isolated plant constituents, the reader is reminded of the phenomenon of synergy – where the interaction of two or more agents results in a combined effect that is greater than the sum of the individual parts (i.e. of the additive effects). The traditional application of this concept is in the methods of combining herbal medicines in formulas; however, in recent times it has also been applied much more broadly, including for the combined effects of active constituents within the same herb. One of the most often referred to example in herbal medicine circles is for the antidepressant effects of St John's wort (*Hypericum perforatum*, Hypericaceae) due to a combination of constituents (Simmen *et al.*, 2001).

The method for testing the occurrence of synergism between chemical compounds was first developed by Loewe in the 1920s and published in German language journals (Tallarida, 2012). The method is based on the production of graphs containing isoboles — lines joining points representing dose combinations of equal effect. These graphs became known as isobolograms or Loewe's synergy.

Table 1.2. Examples of synergistic interactions demonstrated for herbal medicines and essential oils

Herbs and compounds	Synergistic effects	References
Ginkgolides A and B	Inhibition of platelet aggregation	Williamson (2001)
Extracts of *Hydrastis canadensis* leaf and root	Enhanced antimicrobial action of berberine	Ettefagh *et al.* (2011)
Artemisinin and flavonoids	Antimalarial activity	Phillipson (2001), Rasoanaivo *et al.* (2011)
Cinchona alkaloids	Antimalarial activity	Rasoanaivo *et al.* (2011)
Essential oil components in *Ocimum gratissimum*	Anticonvulsive activity	Galindo *et al.* (2010)
Flavonoids in *Ocimum sanctum*	Antibacterial activity	Ali and Dixit (2012)
Reishi, shiitake and maitake mushrooms	Immunomodulatory activity	Mallard *et al.* (2019)
Astragalus root and rehmannia root	Diabetic wound healing	Lau *et al.* (2012)
Carvracol and 1,8-cineole – constituents of essential oils	Controlling bacterial growth in processed vegetables	de Sousa *et al.* (2012), de Oliveira *et al.* (2015)
Murraya koenigii and *Coriandrum sativum* (edible)	Antibacterial multidrug resistant bacteria	Pandey *et al.* (2012)
Leptospermum petersonii and *Leptospermum scoparium* essential oils	Antibacterial against *Staphylococcus aureus* and *Pseudomonas aeruginosa*	Van Vuuren *et al.* (2014)

More recently a new generation of scientists have further streamlined the isobologram method, exploring new methods for qualifying and quantifying the processes involved (e.g. Berenbaum, 1977; Spelman *et al.*, 1999; Williamson, 2001; Tallarida, 2001, 2012; Wagner and Ulrich-Merzenich, 2009; Panossian *et al.*, 2018). Table 1.2 provides examples of synergistic interactions that have been demonstrated for herbal medicines and essential oils.

Note

[1] Chiral comes from the Greek word for hand – it refers to the property whereby the right hand is a mirror image of the left hand.

Recommended reading

Bisset, N.G. (ed.) (1994) *Max Wichtl Herbal Drugs and Phytopharmaceuticals.* Medpharm Scientific Publications, Stuttgart, Germany.

Dewick, P.M. (2009) *Medicinal Natural Products: a Biosynthetic Approach*, 3rd edn. Wiley, Chichester, UK. doi: 10.1002/9780470742761

Phillipson, J.D. (2001) Phytochemistry and medicinal plants. *Phytochemistry* 56(3), 237–243. https://doi.org/10.1016/S0031-9422(00)00456-8

Tucker, A. and Debaggio, T. (2009) *The Encyclopedia of Herbs.* Timber Press, Oregon.

Van Wyk, B. (2013) *Culinary Herbs and Spices of the World.* Kew Publishing, Richmond, UK.

Van Wyk, B. and Wink, M. (2014) *Phytomedicines, Herbal Drugs and Poisons.* Kew Publishing, Richmond, UK.

Wink, M. (2015) Modes of action of herbal medicines and plant secondary metabolites. *Medicines* 2, 251–286. doi: 10.3390/medicines2030251

Wink, M. and van Wyk, B. (2008) *Mind-altering and Poisonous Plants of the World.* Briza Publications, Pretoria, South Africa.

Wink, M. and van Wyk, B. (2017) *Medicinal Plants of the World.* CAB International, Wallingford, UK.

References

Ali, H. and Dixit, S. (2012) *In vitro* antimicrobial activity of flavanoids of *Ocimum sanctum* with synergistic effect of their combined form. *Asian Pacific Journal of Tropical Disease* 2(1), S396–S398. https://doi.org/10.1016/S2222-1808(12)60189-3

Berenbaum, M.C. (1977) Synergy, additivism and antagonism in immunosuppression. A critical review. *Clinical and Experimental Immunology* 28(1), 1–18. Available at: https://www.ncbi.nlm.nih.gov/pmc/articles/PMC1540873/ (accessed 11 March 2020).

Black, N. (1996) Why we need observational studies to evaluate the effectiveness of health care. *British Medical Journal* 312, 1215–1218.

Bloom, B.S., Retbi, A., Dahan, S. and Jonsson, E. (2000) Evaluation of randomised controlled trials on complementary and alternative medicine. *International Journal of Technological Assessment of Health Care* 16(1), 13–21. doi: 10.1017/s0266462300016123

Cronquist, A. (1988) *The Evolution and Classification of Flowering Plants*, 2nd edn. New York Botanical Gardens, New York.

de Oliveira, K.Á.R., de Sousa, J.P., Medeiros, J., de Figueiredo, R.C.B., Magnani, M., et al. (2015) Synergistic inhibition of bacteria associated with minimally processed vegetables in mixed culture by carvacrol and 1,8-cineole. *Food Control* 47, 334–339. doi: 10.1016/j.foodcont.2014.07.014

de Sousa, J.P., de Azerêdo, G.A., de Araújo Torres, R., da Silva Vasconcelos, M.A., da Conceição, M.L. and de Souza, E.L. (2012) Synergies of carvacrol and 1,8-cineole to inhibit bacteria associated with minimally processed vegetables. *International Journal of Food Microbiology* 154(3), 145–151. https://doi.org/10.1016/j.ijfoodmicro.2011.12.026

Ettefagh, K.A., Burns, J.T., Junio, H.A., Kaatz, G.W. and Cech, N.B. (2011) Goldenseal (*Hydrastis canadensis* L.) extracts synergistically enhance the antibacterial activity of berberine via efflux pump inhibition. *Planta Medica* 77(8), 835–840. https://doi.org/10.1055/s-0030-1250606

Fischer, F.H., Lewith, G., Witt, C.M., Linde, K., von Ammon, K., *et al.* (2014) High prevalence but limited evidence in complementary and alternative medicine: guidelines for future research. *BMC Complementary and Alternative Medicine* 14, 46. http://www.biomedcentral.com/1472-6882/14/46

Galindo, L.A., Pultrini, A. and Costa, M. (2010) Biological effects of *Ocimum gratissimum* L. are due to synergic action among multiple compounds present in essential oil. *Journal of Natural Medicines* 64(4), 436–441. doi: 10.1007/s11418-010-0429-2

Jensen, H.D., Struve, C., Christensen, S.B. and Krogfelt, K.A. (2017) Cranberry juice and combinations of its organic acids are effective against experimental urinary tract infection. *Frontiers in Microbiology* 8, 542. https://doi.org/10.3389/fmicb.2017.00542

Kabera, J.N., Semana, E., Mussa, A.R. and He, X. (2014) Plant secondary metabolites: biosynthesis, classification, function and pharmacological properties. *Journal of Pharmacy and Pharmacology* 2(7), 377–392.

Kaur, R., Taheam, N., Sharma, A.K. and Kharb, R. (2013) Important advances on antiviral profile of chromone derivatives. *Research Journal of Pharmaceutical, Biological and Chemical Sciences* (4)2, 79–96. Available at: https://www.rjpbcs.com/pdf/2013_4(2)/[9].pdf (accessed 11 March 2020).

Lau, K.M., Lai, K.K., Liu, C.L., Chor-Wing Tam, J., To, M. -H., *et al.* (2012) Synergistic interaction between Astragali Radix and Rehmanniae Radix in a Chinese herbal formula to promote diabetic wound healing. *Journal of Ethnopharmacology* 141(1), 250–256. doi: 10.1016/j.jep.2012.02.025

Mallard, B., Leach, D.N., Wohlmuth, H. and Tiralongo, J. (2019) Synergistic immuno-modulatory activity in human macrophages of a medicinal mushroom formulation consisting of reishi, shiitake and maitake. *PLoS ONE* 14(11): e0224740. https://doi.org/10.1371/journal.pone.0224740

Pandey, A., Verma, O.P., Nageswaran, N. and Gupta, R. (2012) Synergistic antimicrobial activity of edible herbs against multi drug resistant bacteria. *The Bioscan* 7(1), 103–105.

Panossian, A., Seo, E.-J. and Efferth, T. (2018) Synergy assessments of plant extracts used in the treatment of stress and aging-related disorders. *Synergy* 7, 39–48. https://doi.org/10.1016/j.synres.2018.10.001

Perrine, D.M. (1996) *The Chemistry of Mind Altering Drugs*. American Chemical Society, Washington, DC.

Phillipson, J.D. (2001) Phytochemistry and medicinal plants. *Phytochemistry* 56(3), 237–243. https://doi.org/10.1016/S0031-9422(00)00456-8

Rasoanaivo, P., Wright, C.W., Willcox, M.L. and Gilbert, B. (2011) Whole plant extracts versus single compounds for the treatment of malaria: synergy and positive interactions. *Malaria Journal* 10(Suppl. 1), S4. https://doi.org/10.1186/1475-2875-10-S1-S4

Rauter, A.P., Ennis, M., Hellwich, K.-H., Herold, B.J., Horton, D., *et al.* (2018) Nomenclature of flavonoids (IUPAC Recommendations 2017). *Pure and Applied Chemistry* 90(9), 1429–1486. doi: 10.1515/pac-2013-0919

Sarda, V., Farooqi, J.A. and Malhotra, S. (2017) Basic Organic Chemistry, Unit 13. In: *Chemistry of Elements, Block-5*. Indira Gandhi National Open University (IGNOU) Self-Learning Material (SLM), eGyankosh, National Digital Repository. Available at: http://egyankosh.ac.in/handle/123456789/7583 (accessed 8 March 2020).

Simmen, U., Higelin, J., Berger-Büter, K., Schaffner, W. and Lundstrom, K. (2001) Neurochemical studies with St. John's wort *in vitro*. *Pharmacopsychiatry* 34(Suppl1), 137–142. doi: 10.1055/s-2001-15475

Spelman, K., Duke, J.A. and Bogenschutz-Godwin, M.J. (1999) The synergy principle at work in plants, pathogens, insects, herbivores, and humans. In: Kaufman, P.B., Cseke, L.J., Warber, S., Duke, J.A. and Brielmann, H.L. (eds) *Natural Products from Plants*. CRC Press, Boca Raton, Florida, pp. 475–501.

Tallarida, R.J. (2001) Drug synergism: its detection and applications. *Journal of Pharmacology and Experimental Therapeutics* 298(3), 865–872. Available at: http://jpet.aspetjournals.org/content/298/3/865.long (accessed 12 March 2020).

Tallarida, R.J. (2012) Revisiting the isobole and related quantitative methods for assessing drug synergism. *The Journal of Pharmacology and Experimental Therapeutics* 342(1), 2–8. http://dx.doi.org/10.1124/jpet.112.193474

Van Vuuren, S.S., Docrat, Y., Kamatou, G.P.P. and Viljoen, A.M. (2014) Essential oil composition and antimicrobial interactions of understudied tea tree species. *South African Journal of Botany* 92, 7–14. https://doi.org/10.1016/j.sajb.2014.01.005

Vincent, C. and Furnham, A. (1999) Complementary medicine: state of the evidence. *Journal of Royal Medical Society* 92(4), 170–177. https://doi.org/10.1177/014107689909200403

Wagner, H. and Ulrich-Merzenich, G. (2009) Synergy research: approaching a new generation of phytopharmaceuticals. *Journal of Natural Remedies* 9(2), 121–141.

Waterman, P. and Mole, S. (1994) *Analysis of Phenolic Plant Metabolites*. Blackwell Scientific Publishing, Oxford, UK.

Williamson, E.M. (2001) Synergy and other interactions in phytomedicines. *Phytomedicine* 8(5), 401–409. https://doi.org/10.1078/0944-7113-00060

Wink, M. (2015) Modes of action of herbal medicines and plant secondary metabolites. *Medicines* 2, 251–286. doi: 10.3390/medicines2030251

Wright, M. (2004) *An Introduction to Chinese Herbal Medicine*. Greenbank Publications, Edinburgh, UK.

Zhang, Q., Sharan, A., Espinosa, S.A., Gallego-Perez, D. and Weeks, J. (2019) The path toward integration of traditional and complementary medicine into health systems globally: the World Health Organization Report on the Implementation of the 2014–2023 Strategy. *The Journal of Alternative and Complementary Medicine* 25(9), 869–871. https://doi.org/10.1089/acm.2019.29077.jjw

Phenols

2

Phenols are one of the largest groups of secondary plant constituents. They are aromatic alcohols since the hydroxyl group is always attached to a benzene ring (see diagram below). Like all alcohols, the names of phenols almost always end in the letters 'ol'. In addition, the ring system may bear other substitutes, such as methyl groups and carboxylic acid.

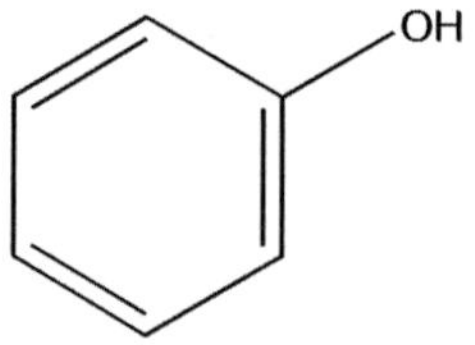

Most phenolic compounds are derivatives of the shikimic acid pathway, though some more complex phenols contain atoms with other biosynthetic origins such as the acetate pathway. Shikimic acid itself is an intermediate metabolite along the shikimic acid pathway, and it rarely occurs in plants, one exception being star anise (*Illicium verum*, Schisandraceae) (van Wyk and Wink, 2017). This molecule is the source of the antiviral drug, oseltamivir phosphate (Tamiflu), and despite efforts to synthesize the drug, it is still derived from seeds of star anise (Ghosh *et al.*, 2012). Shikimic acid in other plants undergoes dehydration and reduction through several stages to form quinic acid, a more stable molecule, which acts as a precursor for a variety of compounds, including the alkaloid quinine. Alternative enzymatic pathways from shikimic acid lead to the aromatic molecules protocatechuic acid and gallic acid, and (with several more steps) to 4-benzoic acid. Structurally these compounds are phenolic acids, and precursors to tannins and other polyphenols (see Chapter 3, this volume).

DOI: 10.1079/9781789243079.0002

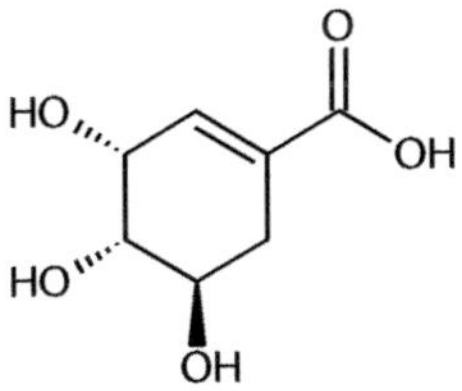

shikimic acid 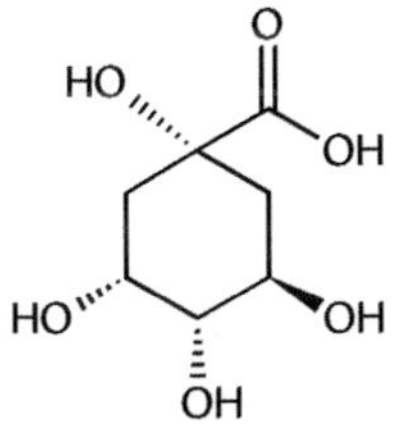   quinic acid 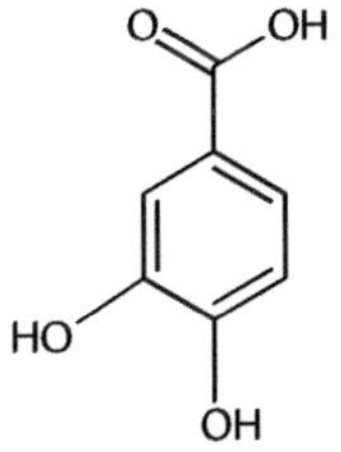   protocatechuic acid

Simple phenols

Simple phenols consist of an aromatic ring in which a hydrogen is replaced by a hydroxyl group, often with additional hydroxyl groups attached to the ring. Their distribution is widespread among all classes of plants. General properties of simple phenols are bactericidal, antiseptic and anthelmintic. Phenol itself is a standard for other antimicrobial agents.

The simplest phenols are six carbon (C_6) structures consisting of an aromatic ring with hydroxyl groups attached. These include **pyrogallol** and **hydroquinone**.

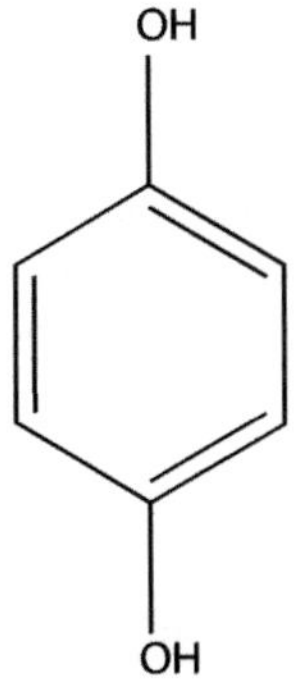

hydroquinone

Addition of a carboxyl group to the basic phenol structure produces a group of C_6C_1 compounds, including some of widespread distribution among plants and with important therapeutic activity. The most important of these are gallic acid and salicylic acid.

A rarer type of simple phenols with C_6C_2 structures are known as acetophenones. These are found in clove buds (*Syzygium aromaticum*, Myrtaceae) (Ryu *et al.*, 2016). Some acetophenones have demonstrated antiasthmatic activity, particularly **apocynin** and its glycoside, **androsin**, which are derived from *Picrorhiza kurroa* (Plantaginaceae) (Dorsch *et al.*, 1994). Apocynin also has anti-inflammatory effects due to inhibition of superoxide-generating NADPH oxidase enzyme (Chandasana *et al.*, 2015), while recent investigations provide promising results for neuroprotection (Feng *et al.*, 2017; Okamura *et al.*, 2018).

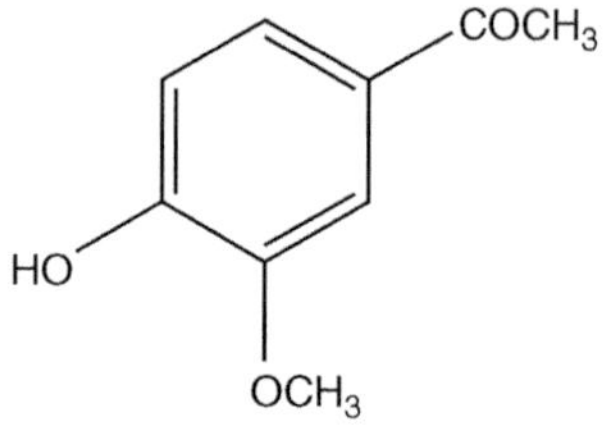

apocynin – an acetophenone

The simple phenol hydroquinone (see above) is derived from hydroxybenzoic acid. Upon glucosylation, **arbutin**, a simple phenol glycoside, is formed. Arbutin occurs in leaves of the pear tree (*Pyrus communis*, Rosaceae) and bearberry (*Arctostaphylos uva-ursi*, Ericaceae) and it is a urinary tract antiseptic and diuretic. Arbutin is metabolized to hydroquinone and conjugates with less than 0.5% free hydroquinone excreted in the urine (Siegers *et al.*, 2003). *A. uva-ursi* is clinically indicated for urinary tract infections, notably cystitis, urethritis and prostatitis. While some authors have noted concerns about the potential toxicity of hydroquinone, a well-known skin-whitening agent, however, de Aribba *et al.* (2013) in their experimental safety evaluation of both the herb and hydroquinone assert there is no evidence linking the herb with any of the purported toxicity, a finding further supported *in vivo* in a toxicity evaluation using rabbits (Saeed *et al.*, 2014). The often cited need to alkalize the urine when using this herb for treating urinary tract infections has also been disproved (Siegers *et al.*, 2003).

Tyrosol and **hydroxytyrosol** (HT) are phenyl ethyl alcohols found in olive oil and white wine. HT, in particular, has been found to contribute to the cardiovascular benefits of olive oil consumption, however, even more promising are the findings with respect to cancer prevention and treatment (Reboredo-Rodríguez *et al.*, 2018).

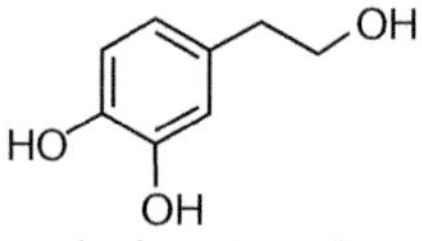

hydroxytyrosol

Amino acids

Another key intermediate along the shikimic acid pathway is **chorismic acid**, amination of which leads to *p*-aminobenzoic acid, better known as the sunscreen molecule PABA, and also occurring as a part of the structure for folic acid. Alternative reactions from chorismic acid lead to the aromatic compounds anthrilic acid (an intermediate to the indole-containing aromatic amino acid L-tryptophan) and prephenic acid (a precursor to both L-tyrosine and L-phenylalanine). Here it can be observed that the shikimic acid pathway may lead to both primary metabolites (amino acids, vitamins) as well as numerous secondary metabolites.

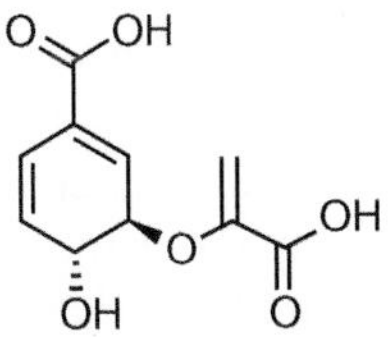

chorismic acid 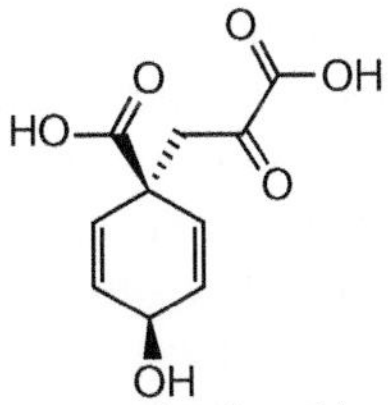 prephenic acid 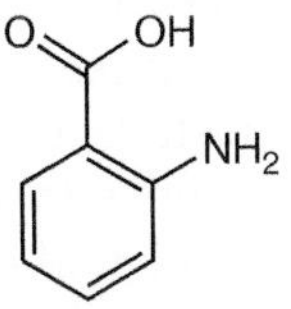 anthrilic acid 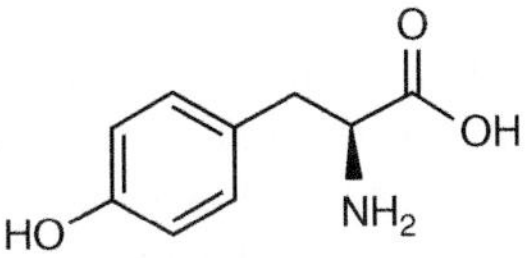 L-tyrosine

Phenylpropanoids

These are C_6C_3 compounds, made up of a benzene ring with a three-carbon side chain. They are derived via elimination of the ammonia side chain of tyrosine to get 4-coumaric acid, and of phenylalanine to obtain cinnamic acid, respectively. Subsequent hydroxylation and methylation reactions produce hydroxycinnamic acids, notably **caffeic**, ***p*-coumaric**, **ferulic** and **sinapic** acids, the precursors to a host of active plant constituents, some of which are discussed below.

Caffeic acid itself is an inhibitor of the enzymes DOPA-decarboxylase and 5-lipoxygenase. It is an analgesic and anti-inflammatory, and it promotes intestinal motility (Adzek and Camarasa, 1988). Caffeic acid is of widespread occurrence, found in green and roasted coffee beans (*Coffea* spp., Rubiaceae).

Hydroxycinnamic acids are widely distributed in plants, both in free form and in the form of esters such as **chlorogenic** and **rosmarinic acids**, two of the most potent plant-derived antioxidant molecules. Note that rosmarinic acid contains an additional benzene ring.

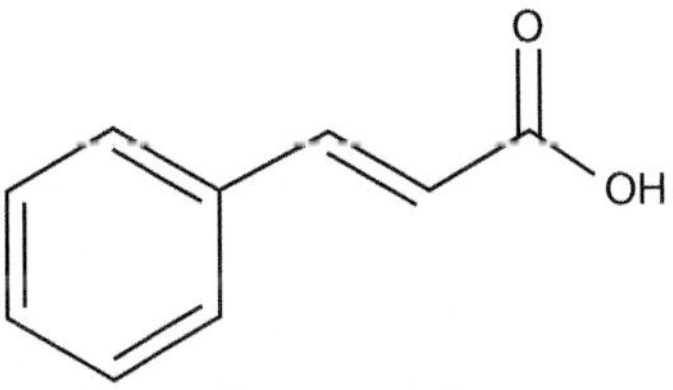

cinnamic acid 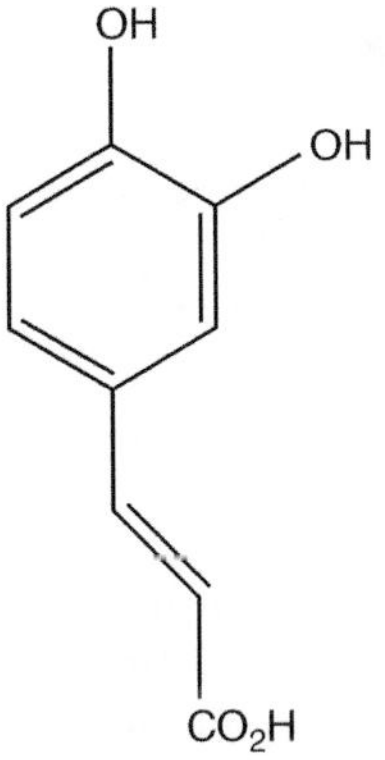 caffeic acid

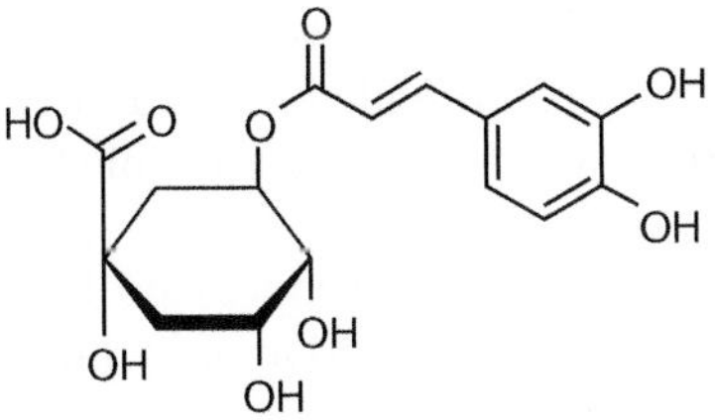

chlorogenic acid 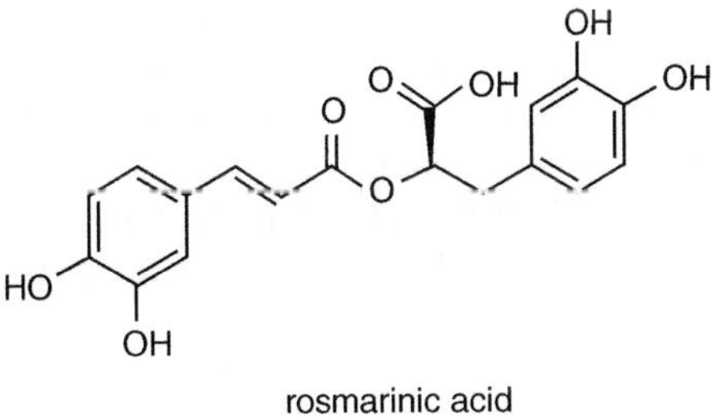 rosmarinic acid

Rosmarinic acid, originally isolated from rosemary (*Salvia rosmarinus* syn. *Rosmarinus officinalis*) in the 1950s, is widely distributed within the Lamiaceae family, notably in sage (*Salvia officinalis*). As a neuroprotective, rosmarinic acid and its derivatives have been studied as a potential treatment for Alzheimer's disease. One derivative, a dimer (two linked rosmarinic acid molecules) known as **salvionolic acid** and found in the Chinese sage (*Salvia miltiorrhiza*, Lamiaceae), is particularly promising, *in vivo* studies demonstrating effects on amyloid β and tau proteins, as well as aiding recovery of neuronal loss (Habtemariam, 2018).

Cinnamic acids are of much benefit therapeutically and are non-toxic. They may also occur as glycosides and depsides such as **cynarin**, a dicaffeoylquinic acid derivative from globe artichoke (*Cynara scolymus*, Asteraceae). Cynarin is formed from the bonding of caffeic and quinic acids. Cynarin and related caffeoylquinic acid isomers are responsible for many of the therapeutic benefits linked to globe artichoke leaf, including as treatment for liver toxicity, metabolic disorders, dyspepsia and hypercholesterolaemia (Lattanzioa *et al.*, 2009; Salem *et al.*, 2017).

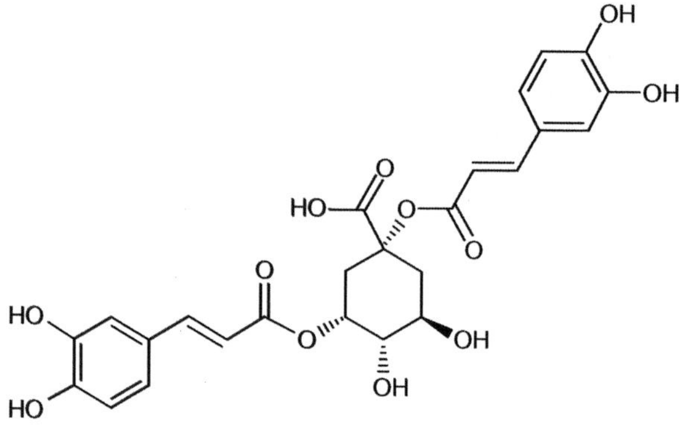

1,3-O-dicaffeoylquinic acid (cynarin)

Ten caffeoylquinic acid derivatives have been identified in artichoke leaf, consisting of four mono-caffeoylquinic and six di-O-caffeoylquinic acid isomers, dependent on whether they contain one or two benzene rings. The most abundant of these in artichoke leaf extract is the monomer chlorogenic acid (39%) while cynarin is but a minor component (1.5%) (Lattanzioa *et al.*, 2009). This suggests that while cynarin is regarded as the signature component within this category, it is more likely the additive or synergistic effects of all ten of these compounds that are responsible for therapeutic efficacy.

Modification of the side chain of hydroxycinnamic acids produces alcohols such as coniferyl alcohol, which act as precursors to the formation of lignin, a high-molecular-weight polymer that gives strength and structure to stems of herbs and the trunks of trees.

By modification of their C_3 side chains or changes in substitution patterns of their aromatic nucleus, hydroxycinnamic acids can form various secondary compounds including phenolic ethers, lignans, coumarins, glycosides, and dimers such as curcumin from the roots of turmeric.

Curcumin (diferuloylmethane) is the yellow pigment from the turmeric rhizome (*Curcuma longa*, Zingiberaceae). Curcumin and its derivatives, referred to as curcuminoids, contain diarylheptanoid structures. Total curcuminoids make up 1–6% of an extract of turmeric rhizome, and these consist of 60–70% by weight of curcumin, 20–27% demethoxycurcumin and 10–15% bisdemethoxycurcumin (Nelson *et al.*, 2017).

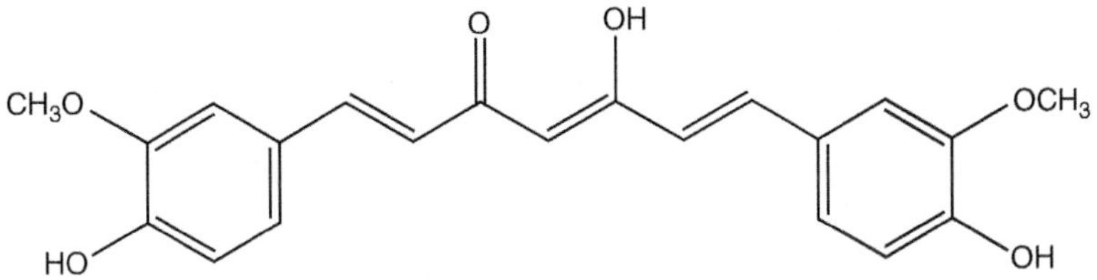

curcumin – diferuloylmethane

Curcumin is insoluble in water but somewhat soluble in methanol and more so in other organic solvents. In solution it exists as a tautomeric enol, that is, one or other of the keto groups converted to an alcohol (Hatcher *et al.*, 2008).

It is well documented that bioavailability of curcumin is poor due to rapid biotransformation in the liver. Curcumin metabolism can be slowed by the addition of black pepper (*Piper nigrum*, Piperaceae) since piperine (an alkaloid in black pepper) is a known inhibitor of hepatic and intestinal glucuronidation (Nawaz *et al.*, 2011). In addition, some of the metabolic or breakdown products of curcumin have been shown to have similar or equivalent actions *in vitro*. These include anti-inflammatory, antitumour, wound healing and cardioprotective activity (Hatcher *et al.*, 2008; Nawaz *et al.*, 2011).

Not all of the benefits attributed to curcumin have been verified by human studies, leading to concern by some observers that – like some other polyphenol compounds – curcumin acts as both a PAINS (pan-assay interference compounds) and an IMPs (invalid metabolic panaceas) candidate (Nelson *et al.*, 2017). These compounds tend to react with various items in biological assays such as proteins, metals, membranes and fluorescence, thereby often providing false positives (i.e. beneficial activities that aren't reproducible in humans). Despite such concerns, curcumin has demonstrated some positive outcomes in clinical studies, including for joint arthritis (Daily *et al.*, 2016), inflammatory bowel disease (Mazieiro *et al.*, 2018) and major depressive disorders (Al-Karawi *et al.*, 2015).

Salicylates and salicins

Salicylic acid is a carboxylated phenol, that is, a carboxylic acid and a hydroxyl group added to a benzene ring. Such a basic molecule can be synthesized in at least three different ways, the simplest being the hydroxylation of benzoic acid (Dewick, 2009). The acid form rarely occurs freely in plants, but usually occurs as glycosides (e.g. **salicin, salicortin**), esters and salts. The plant mainly associated with salicins is willow bark, from several species in the *Salix* genus (Salicaceae). However, the compound, salicin is fairly widespread in other plants, and in one study the level of salicins detected in poplar bark (*Populus* spp., in the same family, Salicaceae, as willow) was far higher than was found in several samples of willow bark (Luo *et al.*, 1998). Unfortunately, the actual species of poplar bark tested in that study was not cited. Another study found wintergreen (*Gaultheria procumbens*, Ericaceae) to be the highest known plant source of salicylates at 10 mg/g fresh weight (Ribnicky *et al.*, 2003).

In humans salicin and related glycosides are first hydrolysed to the aglycone, salicyl alcohol, with the aid of intestinal bacteria. Upon oxidation in the liver and bloodstream salicylic acid is produced (Mahdi, 2010). Salicylic acid undergoes hepatic biotransformation, and most is excreted in the urine as salicylic acid conjugates. **Aspirin** (acetylsalicylic acid) is a synthetic derivative of salicylic acid. Salicin-containing herbs such as willow and poplar barks are primarily used as analgesics, anti-inflammatories and febrifuges. Since they lack the acetyl group found in aspirin, natural salicylates do not have antiplatelet (blood thinning) effects (Meier and Liebi, 1990). The well-documented tendency to gastric haemorrhage associated with aspirin is not a problem in salicin-containing herbs (Bissett, 1994; van Wyk and Wink, 2017). However, caution is required in using them in individuals with aspirin or salicylate sensitivity.

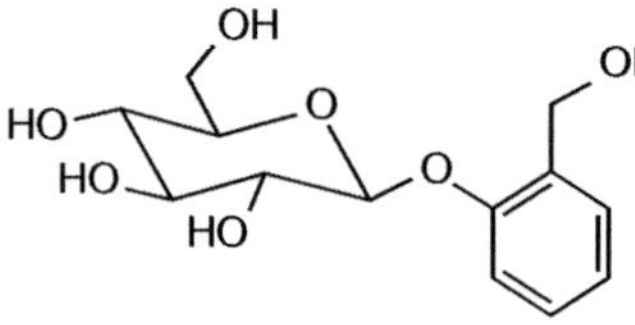

salicin

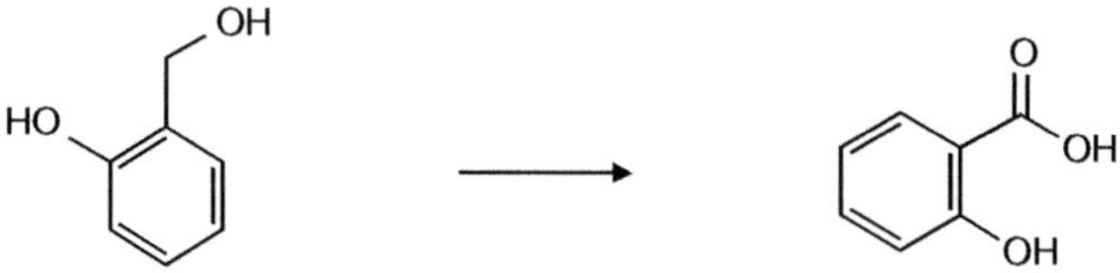

salicyl alcohol   salicylic acid

The structure of salicylic acid was first resolved by the Italian chemist Piria, in 1838. It was later synthesized by the German chemist Kolbe in 1860, following which it was scaled up for industrial production (Mahdi, 2010). The subsequent synthesis of acetylsalicylic acid in 1899 by the Bayer Company resulted in aspirin.

Some derivatives of salicylic acid

Glycosides
- **populin** – poplar bark (*Populus* spp., Salicaceae);
- **gaultherin** – wintergreen (*Gaultheria procumbens*, Ericaceae); and
- **spiraein** – meadowsweet (*Filipendula ulmaria*, Rosaceae).

Esters
- **methyl salicylate** – found in *Filipendula* (Rosaceae) and *Gaultheria* spp. (Ericaceae);
- **salicylaldehyde** – meadowsweet; and
- **acetylsalicylic acid** (aspirin) – aspirin is a synthetic derivative of salicylic acid.

Pharmacological actions of salicylates and their derivatives

Aspirin blocks synthesis of prostaglandins through acetylation of the enzyme cyclooxygenase. Whereas cyclooxygenase-2 (COX-2) is found mainly in inflamed tissues, cyclooxygenase-1 (COX-1) is present in platelets. The inhibition of COX-1 by aspirin is behind the well-known blood-thinning effects of the drug, as well as other adverse reactions such as haemorrhage and gastric irritation.

Other actions associated with salicylic acid derivatives include central nervous system depression and antipyretic effects – they act to increase peripheral blood flow and sweat production, by direct action on the thermogenic section in the hypothalamus. This helps explain the use of salicin-containing herbs for neuralgias, sciatica, myalgia and headaches.

Clinical investigations in Germany based around high-dose willow bark extracts (120–240 mg salicin daily), demonstrated significant reduction of back pain with few side effects (Chrubasik *et al.*, 2000). Also of note, a recent analytical study of *Salix* spp. revealed the presence of a novel salicin derivative salicin-7-sulfate in some species including *Salix alba* (Noleto-Dias *et al.*, 2018). The authors hypothesize this sulfur-containing molecule could increase the anticoagulant and potential for digestive irritant effects in those species that contain it.

Lignans

Lignans are dimeric compounds such as pinoresinol, in which phenylpropane (C_6C_3) units are coupled at the central carbon (usually C-8) of their side chains, to form three-dimensional networks.

Nordihydroguaiaretic acid (NDGA) also called masoprocol from chaparral aka the creosote bush (*Larrea tridentata*, Zygophyllaceae) is another example of a simple lignan. Due to the presence of two catechol groups it has potent antioxidant action, shown to be superior to standard reference compounds such as glutathione and uric acid (Lu *et al.*, 2010). However, concerns about nephrotoxicity shown in animal studies when administered at high doses, led to cessation of use of NDGA as an antioxidant and preservative in the food industry. Potential clinical applications for use of NDGA in the treatment of cardiovascular diseases and cancer are still being investigated, while herbalists continue to use topical oils and salves from the creosote bush for keratosis, fungal infections and other skin ailments (Garza, 2015).

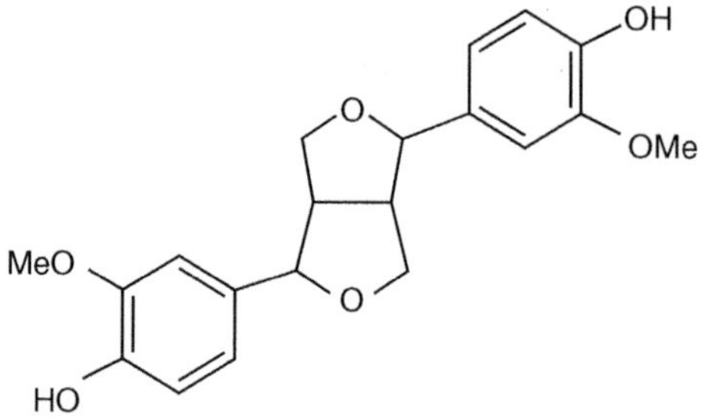

simple lignan: pinoresinol

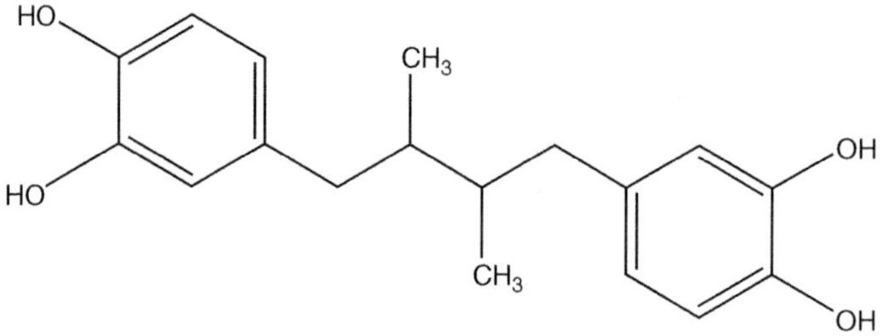

nordihydroguaiaretic acid (NDGA)

Neolignans are similar dimeric structures but their phenylpropane units are coupled differently (Dewick, 2009). Hybrid lignans, or lignoids, are compounds with mixed biosynthetic origin. Examples are flavonolignans such as silybin from *Silybum marianum* (Asteraceae) and xantholignans from *Hypericum perforatum* (Hypericaceae) (Bruneton, 1995).

Fruit and seeds from *Schisandra chinensis* (Schisandraceae) are rich in lignans such as **schizandrin** which is notable for having numerous methoxy groups attached to the main skeleton. In many derived structures such as gomasin A, two adjacent methoxy groups are bonded to form a lactone ring.

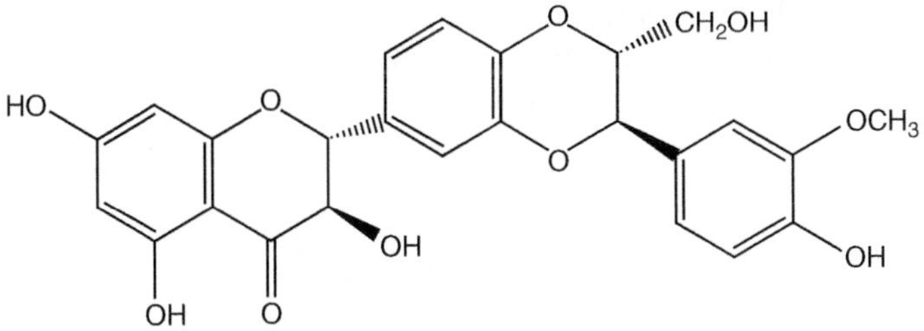

silybin

Schizandrin has been shown to boost mental performance in healthy adults, while **gomasin A** inhibits release of arachidonic acid, and also inhibits leuko-triene B_4 via a separate mechanism, mechanisms likely to be responsible for the anti-inflammatory and antihepatotoxic effects associated with *S. chinensis* (Panossian and Wikman, 2008).

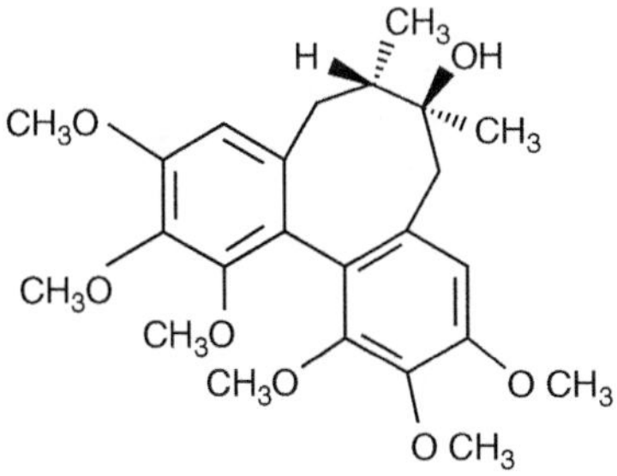

schizandrin

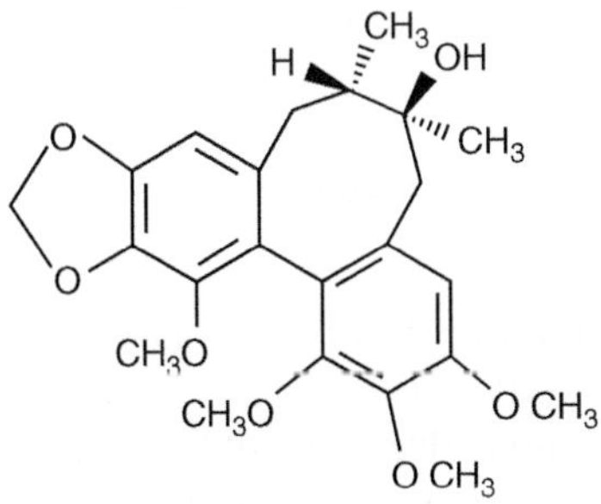

gomasin A

Podophyllotoxin, a neolignan from *Podophyllum* species (Berberidaceae), is an antimitotic, tubulin-binding agent which has been investigated as a source of naturally occurring lead compounds in cancer research (Apers *et al.*, 2003). Etoposide, a semisynthetic derivative of podophyllotoxin glycoside, is used in combination with chemotherapy drugs for treatment of small cell lung cancer, testicular cancer and lymphomas. This compound acts as a topoisomerase II inhibitor, preventing DNA synthesis and replication in cancer cells (Dewick, 2009). The main source of podophyllotoxin comes from the Indian species, *Podophyllum hexandrum*, while the American species, *Podophyllum peltatum* has significantly lower levels of lignans. Podophyllin resin is a concentrated ethanol extract mainly from the Indian species; the resin is used in the form of a paint for treatment of venereal warts.

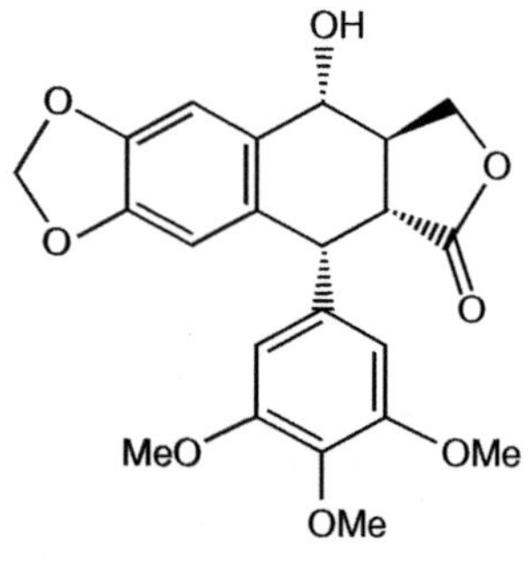

podophyllotoxin

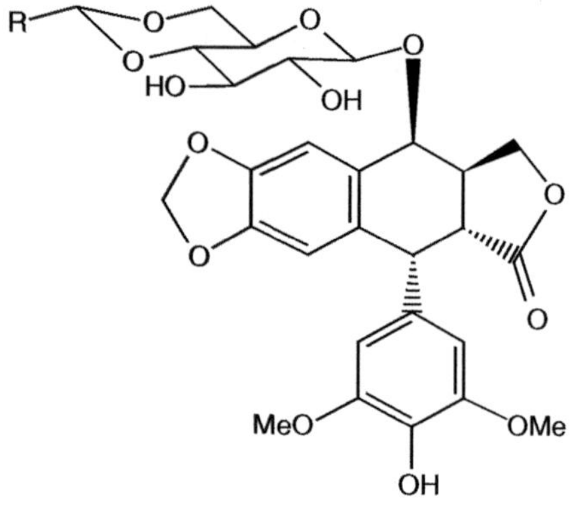

etoposide

Many lignans, including those from flaxseed (*Linum usitatissimum*, Linaceae) and stinging nettle (*Urtica dioica*, Urticaceae) root, are converted by intestinal bacteria to the active metabolites **enterolactone** and **enterodiol**, which are readily absorbed. These two lignans are also metabolites of dietary lignans found in grains and pulses. They have oestrogenic, antitumour and antioxidant properties, and are a major source of the class of compounds known as phytoestrogens.

Coumarins

Coumarins are lactones of hydroxycinnamic acids, with cyclic C_6C_3 skeletons. Coumarin itself is derived from cinnamic acid via *trans–cis* isomerization and lactonization. This molecule is found in many legumes (Fabaceae), such as sweet clover (*Melilotus officinalis*), *Trifolium* spp., as well as in sweet woodruff (*Asperula odorata*, Rubiaceae), and is at least partially responsible for the characteristic 'newly mown hay' aroma. **Dicoumaral** (bishydroxycoumarin), used as an anticoagulant drug, was originally derived from sweet clover, but only occurs when the herb is poorly dried or left out in the open to go mouldy. **Warfarin**, the blood-thinning drug also used as rat poison, is a synthetic dicoumarol derivative.

Coumarins derived from 4-coumaric acid via *trans–cis* isomerization/lactonization, are referred to as simple coumarins and are based on an umbelliferone prototype. Most simple coumarins are substituted with OH or OCH_3 at positions C-6 and C-7. They often occur in glycosidic form, for example, **aesculin** is the glycoside of **aesculetin**.

coumarin

aesculetin (6,7-dihydroxycoumarin)

umbelliferone (7-hydroxycoumarin)

Furanocoumarins have a furan ring at C-6 and C-7 (psoralen) or C-7 and C-8 (angelican) of the coumarin ring system; however, they are not phenolic in structure. **Angelican** and the structurally complex coumarin **archangelican** occur in the roots of angelica (*Angelica archangelica*, Apiaceae) and have spasmolytic activity.

The linear furanocoumarin **psoralen** lends its name to a group of related compounds, psoralens, noted for their phototoxic and photosensitizing properties. These coumarins tend to be concentrated in various species of the Apiaceae and Rutaceae families, including a number of medicinal and food plants such as bishop's weed (*Ammi majus*, Apiaceae), celery (*Apium graveolens*, Apiaceae), bergamot (*Citrus bergamia*, Rutaceae) and rue (*Ruta graveolens*,

Rutaceae). Taking psoralens orally or topically can lead to severe sunburn, blistering and pigmentation of the skin when exposed to sunlight, hence, they should be taken/applied with caution and prolonged sun exposure should be avoided (Gardner and McGuffin, 2013). On the other hand, *A. majus* fruit, particularly rich in furanocoumarins such as xanthotoxin, has been used medically to treat vitiligo, psoriasis and tinea versicolor hypopigmentation, where the subject is concurrently exposed to solar radiation (Al-Snafi, 2013).

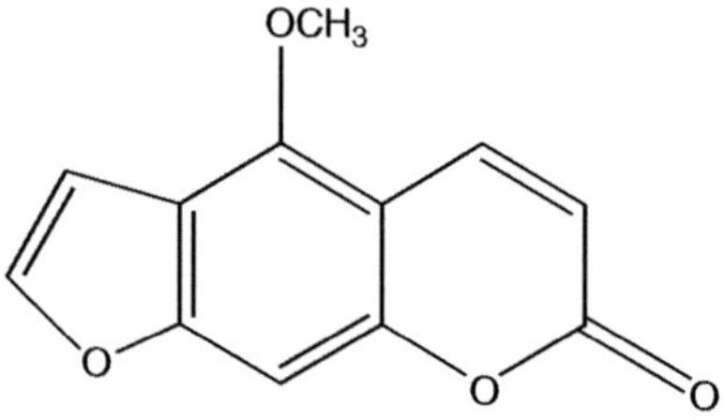

bergapten – a linear furanocoumarin

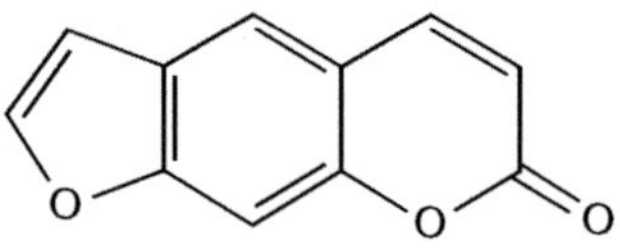

psoralen – a linear furanocoumarin

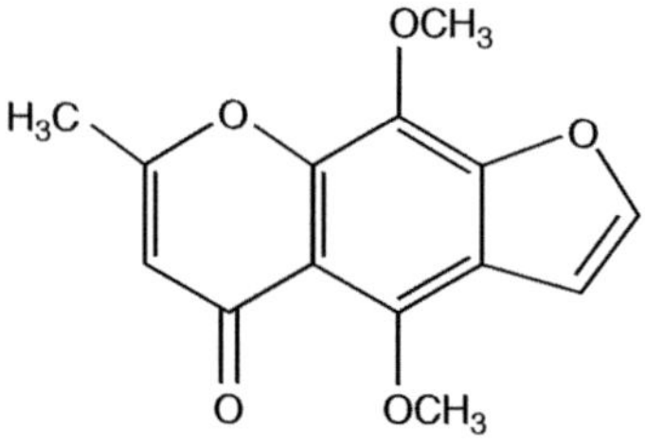

khellin – a furanochromone

Furanochromones are tricyclic compounds related to coumarins, they are also referred to as γ-pyrones. In **khellin** and **visnagin**, active constituents of khella (*Ammi visnaga*, Apiaceae), the main skeleton bears methyl and methyoxy group substitutes. These compounds are recognized vasodilators, bronchodilators and spasmolytic agents, providing scientific rationale for the traditional and contemporary use of khella for treatment of asthma, angina pectoris, hypertension and renal calculi (Günaydin and Beyazit, 2004). **Chromoglycate**, a synthetic derivative of khellin, is a pharmaceutical drug widely prescribed for the treatment of allergic asthma, hay fever and rhinitis. Khellin in conjunction with UV treatment has also been used successfully for treating vitiligo (de Leeuw *et al.*, 2011).

Pyranocoumarins, such as **visnadin**, **samidin** and **dihydrosamidin** containing a pyran ring fused at C-7 and C-8, are also present in *A. visnaga* (Greinwald and Stobernack, 1990; Khalil *et al.*, 2020).

Polyketide-derived phenols

Furanochromones, while often categorized with the coumarins, are in fact polyketide derivatives. When a cinnamic acid ester (cinnamoyl-CoA) combines with acetyl-CoA with an additional side chain of keto groups provided by malonyl-CoA, a polyketide backbone is formed, reflecting origins in both the shikimic acid and acetate pathways (Dewick, 2009).

This complex biosynthetic origin gives rise to a host of polyphenolic compounds, including flavonoids, chalcones and stilbenes, as determined by specific synthase enzymes. For example, when cinnamoyl-CoA undergoes reactions catalysed by stilbene synthase enzyme, the end products are stilbenes.

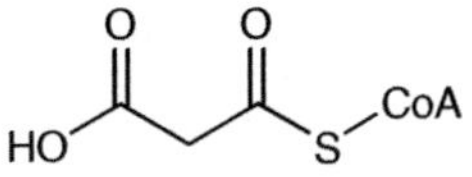

malonyl-CoA

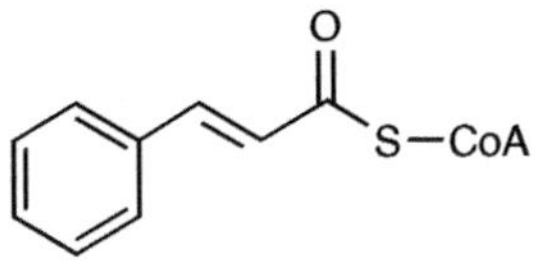

cinnamoyl-CoA

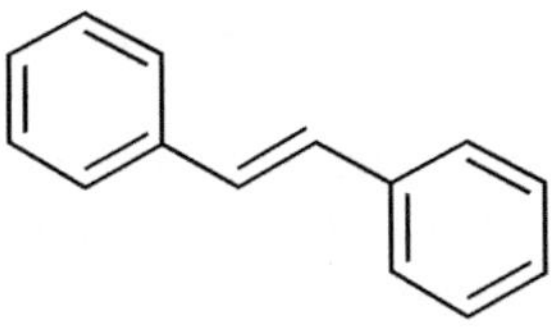

trans-stilbene

Stilbenes

Stilbenes are characterized by two benzene rings, which are derived from cinnamoyl-CoA (described above). These compounds are not widespread in plants, however, **resveratrol** (3,4,5-trihydroxy-*trans*-stilbene) does occur in many plant species, having been initially isolated from the roots of the white hellebore (*Veratrum album*, Melanthiaceae). Most attention has been paid to another source – the skins of grapes (*Vitis vinifera*, Vitaceae) – where it acts as a phytoalexin, compounds synthesized as protective mechanisms against fungal infection and other environmental triggers (Creasy and Creasy, 1998; Park and Pezzuto, 2015). It is present in red wine, being one of the health-promoting polyphenols found in the beverage, however, the main source used as a dietary supplement is Japanese knotweed (*Reynoutria japonica*, Polygonaceae), previously known as *Fallopia japonica* or *Polygonum cuspidatum*. This oriental species is an invasive weed in North America, Europe and New Zealand and hence is in plentiful supply.

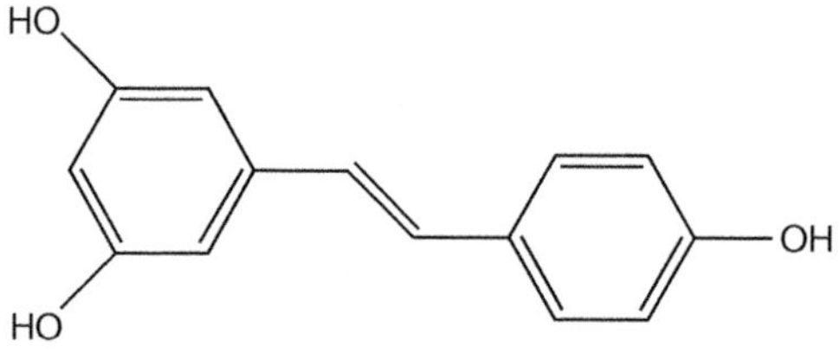

resveratrol – a hydroxystilbene

Resveratrol is most commonly used as an isolate; however, herbalists who prefer using whole plant preparations can utilize either grape seeds or the Japanese knotweed. As of 2014, over 6500 papers have been published on resveratrol, reflecting its status as one of the most widely used dietary supplements in the world (Park and Pezzuto, 2015). Uses for resveratrol are many and varied, being cardioprotective, anti-inflammatory, antioxidant, anti-ageing and as a cancer preventative. Recent research has focused on metabolic disorders and ageing, with the identification of specific molecular targets such as SIRT1 (Sirtuin-1) and AMPK (AMP-activated protein kinase) (Kulkarni and Cantó, 2015).

Quinones

Quinones are another class of compounds that may be formed from either the acetate or shikimic acid pathway, in fact, any of four different metabolic pathways may be involved in quinone biosynthesis (Harborne and Baxter, 1993). Quinones are polycyclic aromatic compounds in which one hexane ring contains two opposite carbonyl groups. Typically, the quinoid structure is attached to one or more benzene rings, which may or may not have a hydroxyl. The simplest quinone, **benzoquinone**, lacks the benzene ring altogether.

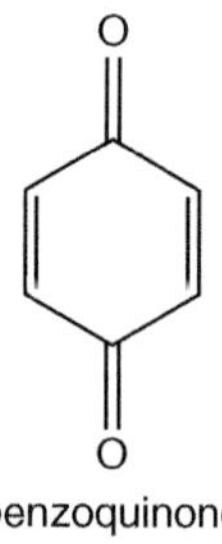

benzoquinone

Quinones form an important component of the electron transport system in plants and mammals. **Ubiquinol**, the reduced form of coenzyme Q10 (CoQ10) and **menaquinone** (vitamin K) have significant antioxidant properties and play a major role in protecting cells from free-radical damage. The largest subgroup is the anthraquinones, which occur mainly as glycosides and are referred to in Chapter 4, this volume. For this chapter we will focus on a smaller group, the naphthoquinones.

Naphthoquinones

Naphthoquinones are characterized by their dark pigmentation. **Lawsone** is the active principle in the popular hair dye obtained from the henna plant (*Lawsonia inermis*, Lythraceae), while **shikonin** is found in the roots of the red dye plants, *Alkanna tinctoria* and *Lithospermum erythrorhizon* (both in the Boraginaceae family). Many 1,4-naphthoquinones – in which oxygen is double-bonded to carbon in the C-1 and C-4 positions in the ring – are recognized for their antimicrobial, antifungal and antitumour activities. These include **juglone** from walnut bark (*Juglans nigra*, *Juglans regia*, Juglandaceae), **lapachol** from pau d'arco (*Handroanthus impetiginosus*, Bignoniaceae; formerly *Tabebuia impetiginosa*) and **plumbagone** from sundew (*Drosera rotundifolia*, Droseraceae). Lapachol has an isoprene unit (five-carbon chain) attached to the quinone ring; juglone has a simple bicyclic structure with a phenolic functional group.

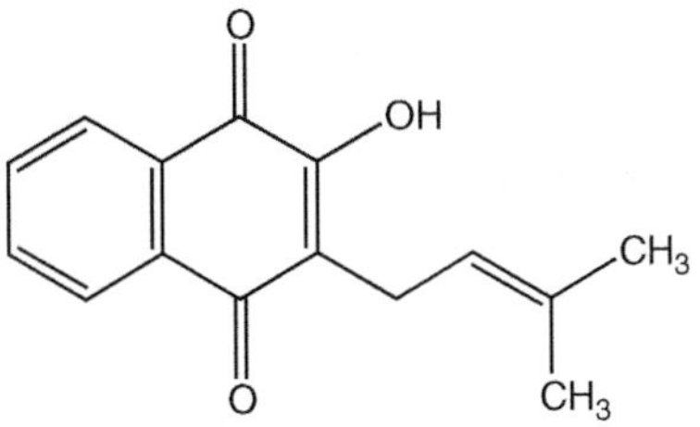

lapachol

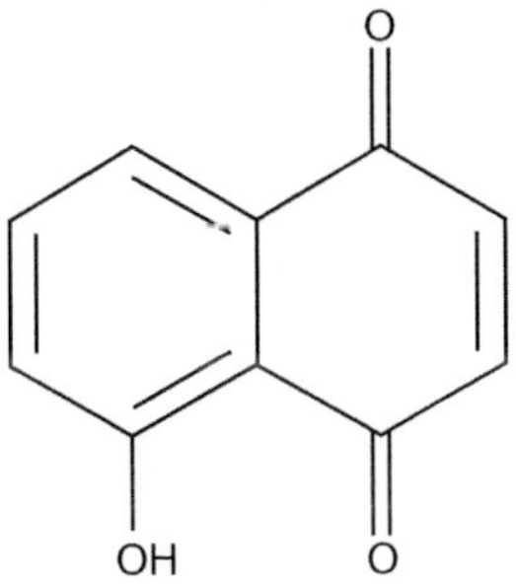

juglone (5-hydroxy-1,4-naphthoquinone)

In 1968, lapachol was identified as an antitumour agent, showing significant activity against Walker-256 carcinosarcoma *in vivo*, following twice daily oral administration (Rao *et al.*, 1968). The isopentenyl side chain in lapachol is thought to play a pivotal role in antitumour activity. Although clinical studies are lacking, experimental research indicates lapachol inhibits a broad range of human pathogens and parasites, the latter including *Schistosoma mansoni*, *Trypanosoma cruzi*, *Leishmania* spp. and the mollusc *Biomphalaria glabrata* (Hussain *et al.*, 2007). Shikonin has also been shown to inhibit the growth of breast cancer cells (Widhalm and Rhodes, 2016).

Another interesting field for naphthoquinone research is in allelopathy, defined by Wikipedia as a 'biological phenomenon by which an organism produces one or more biochemicals that influence the germination, growth, survival, and reproduction of other organisms'. In a classic paper Australian botanist R.J. Willis reviewed the inhibiting effects of the *Juglans* genus, especially *J. nigra*, on the germination and growth of hundreds of species of plants, mainly attributed to the compound juglone (Willis, 2000). Other naphthoquinones-containing species also demonstrate allelopathic properties, including the invasive weed known in Australia as Paterson's curse (*Echium plantagineum*, Boraginaceae), whose roots contain **shikonin**.

Naphthodianthrones – hypericin

Although more closely related to the anthraquinones (see Chapter 4 'Glycosides', this volume), naphthodianthrones are covered here since they are derived from polyketides, and unlike anthraquinones they don't form glycosides. **Hypericin**, the dark-red pigment from *Hypericum perforatum* (Hypericaceae), while being structurally similar to dianthrones such as the sennosides, does not break down to anthrone in the bowel and is without laxative action. Hypericin and **pseudohypericin** have been thoroughly investigated (generally in *Hypericum* extracts standardized to hypericin content) for antidepressant, antiviral, antibacterial, antitumour and anti-inflammatory activities (Mir *et al.*, 2019).

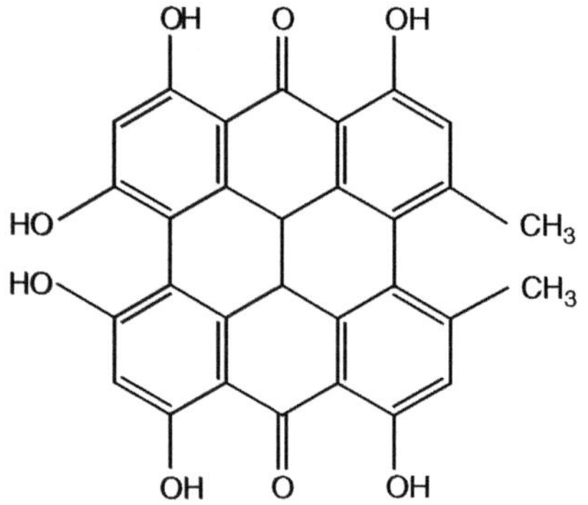

hypericin – a naphthodianthrone

Hypericin and its derivatives accumulate in dark nodular structures, in the aerial parts of *H. perforatum* and other *Hypericum* species. Herbalists are familiar with the red staining of the hands that occurs when harvesting the aerial parts, and infusions in fixed oils or organic solvents display characteristic deep-red colour. A powerful photosensitizing agent, hypericin requires the presence of light to trigger antiviral and other therapeutic effects (Jendželovská *et al.*, 2016). In laboratory investigations, hypericin inhibition of enveloped viruses such as human immunodeficiency virus (HIV), herpes simplex virus (HSV) and influenza A virus increases up to 100 times in the presence of light (Kubin *et al.*, 2005). In recent years hypericin has been investigated for anti-cancer photodynamic therapy (PDT) and photodynamic diagnosis (PDD), and is considered a highly promising fluorescent photosensitizer for both detection and treatment of cancer (Jendželovská *et al.*, 2016). In one *in vivo* study, hypericin PDT therapy was toxic to both pigmented and unpigmented metastatic melanoma cells (Kleemann *et al.*, 2014).

Safety issues with St John's wort
Despite these potential benefits, photosensitization can also cause phototoxic skin reactions, and increase the risk of cataracts. Depending on the level of sun exposure, the amount of St John's wort required to induce phototoxicity in an adult has been calculated at 2–4 g daily, or 5–10 mg of hypericin (Butterwreck, 2009). The other safety concern for St John's wort is for potential drug inter-actions. Hypericin (as well as the unrelated hyperforin group of constituents) are known to interact with some of the cytochrome P450 (CYP450) enzymes responsible for the metabolism of many pharmaceutical drugs, interactions which may either reduce or increase their bioavailability (Kubin *et al.*, 2005).

Kavalactones

Kava kava (*Piper methysticum*, Piperaceae) is a large shrub widely cultivated in Oceania. The dried rhizome and root are traditionally used in preparation of a mildly intoxicating beverage. Kava contains a unique group of polyketide-derived lactones that follow a similar biosynthetic pathway to the chalcones and flavonoids (Pluskal *et al.*, 2019). These are kavain, dehydrokavain, yangonin, demethoxyyangonin, methysticin and dihydromethysticin. Structurally they consist of benzene and lactone rings linked by a C_2 chain. A related group known as flavokavains contain a ketone functional group attached to the carbon linkage (Liu *et al.*, 2018). Dihydrochalcones, cinnamic acid esters and flavonones are also present (Xuan *et al.*, 2008).

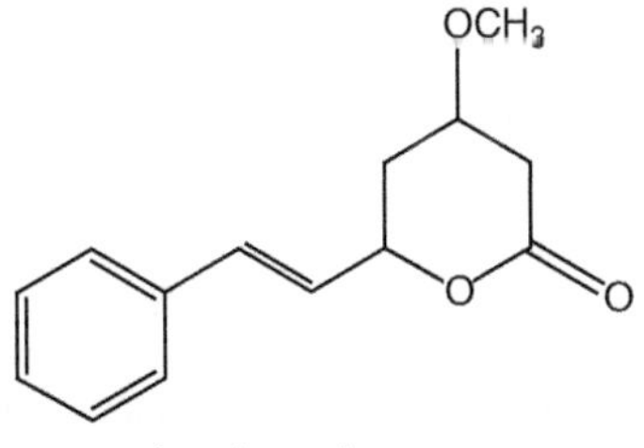

kavain – a kava pyrone

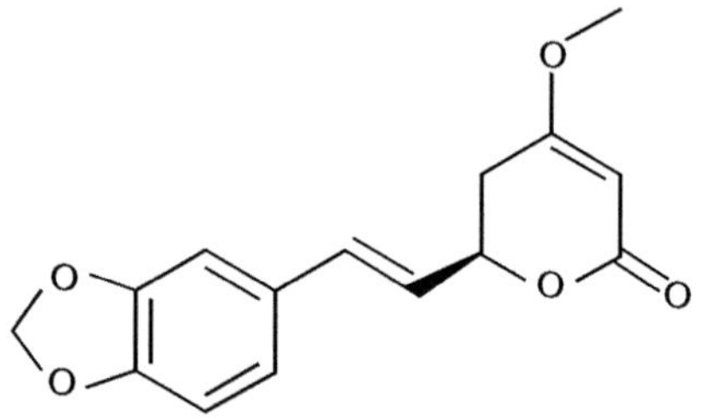

methysticin

Kavalactones are potent, centrally acting skeletal muscle relaxants. They act as hypnotics, antipyretics, sedatives, local anaesthetics, smooth muscle relaxants and antifungal agents. Studies in humans have confirmed kava's reputation for relieving anxiety and depression (Sarris *et al.*, 2009) and general anxiety syndrome (Ooi *et al.*, 2018). These studies were based on aqueous extracts. Some, but not all kava lactones are water soluble. A comparison of solvents used for extracting kavalactones revealed acetone to be the most efficient, however, there are safety issues to consider for this solvent, and it is rarely used for preparation of medicinal extracts (Xuan *et al.*, 2008; Teshcke, 2011).

Safety concerns for kava are mainly centred on the potential for hepatotoxicity, although occurrences are quite rare (Lude *et al.*, 2008; Gardner and McGuffin, 2013).

References

Adzek, T. and Camarasa, J. (1988) Pharmacokinetics of polyphenolic compounds. *Herbs, Spices and Medicinal Plants* 3, 25–47.

Al-Karawi, D., Al Mamoori, D.A. and Tayyar, Y. (2015) The role of curcumin administration in patients with major depressive disorder: mini meta-analysis of clinical trials. *Phytotherapy Research* 30(2), 175–183. https://doi.org/10.1002/ptr.5524

Al-Snafi, A.E. (2013) Chemical constituents and pharmacological activities of *Ammi majus* and *Ammi visnaga*: a review. *International Journal of Pharmacy and Industrial Research* 3(3), 257–265. Available at: https://www.academia.edu/11590562/Chemical_Constituents_and_Pharmacological_Activities_of_Ammi_majus_and_Ammi_visnaga._A_review (accessed 25 November 2020).

Apers, S., Vlietinck, A. and Pieters, L. (2003) Lignans and neolignans as lead compounds. *Phytochemistry Reviews* 2, 201–217. https://doi.org/10.1023/B:PHYT.0000045497.90158.d2

Bissett, N.G. (ed.) (1994) *Herbal Drugs and Phytopharmaceuticals*. Medpharm Publications, Stuttgart, Germany.

Bruneton, J. (1995) *Pharmacognosy, Phytochemistry, Medicinal Plants*. Lavoisier Publications, Paris.

Butterwreck, V. (2009) St. John's wort: quality issues and active compounds. In: Cooper, R. and Kronenberg, F. (eds) *Botanical Medicine: From Bench to Bedside*. Mary Ann Liebert, New York. doi: 10.1089/9781934854051.69

Chandasana, H., Chhonker, S., Bala, V., Prasad, Y.D. and Bhatta, R.S. (2015) Pharmacokinetic, bioavailability, metabolism and plasma protein binding evaluation of NADPH-oxidase inhibitor apocynin using LC–MS/MS. *Journal of Chromatography B*, 985(15), 180–188. doi: 10.1016/j.jchromb.2015.01.025

Chrubasik, S., Eisenberg, E., Balon, E., Weinberger, T., Luzzati, R. and Conradt, C. (2000) Treatment of low back pain exacerbations with willow bark extract: a randomised double-blind study. *American Journal of Medicine* 109(1), 9–14. doi: 10.1016/s0002-9343(00)00442-3

Creasy, L.L. and Creasy, M.T. (1998) Grape chemistry and the significance of resveratrol: an overview. *Pharmaceutical Biology* 36(Supplement), 8–13.

Daily, J.W., Yang, M. and Park, S. (2016) Efficacy of turmeric extracts and curcumin for alleviating the symptoms of joint arthritis: a systematic review and meta-analysis of randomized clinical trials. *Journal of Medicinal Food* 19, 717–729. doi: 10.1089/jmf.2016.3705

de Arriba, S.G., Naser, B. and Nolte, K.U. (2013) Risk assessment of free hydroquinone derived from *Arctostaphylos uva-ursi* folium herbal preparations. *International Journal of Toxicology* 32(6), 442–453. doi: 10.1177/1091581813507721.

de Leeuw, J., Assen, Y.J., van der Beek, N., Bjerring, P. and Martino Neumann, H.A. (2011) Treatment of vitiligo with khellin liposomes, ultraviolet light and blister roof transplantation. *Journal of the European Academy of Dermatology and Venereology* 25(1), 74–81. doi: 10.1111/j.1468-3083.2010.03701.x

Dewick, P.M. (2009) The shikimate pathway: aromatic amino acids and phenylpropanoids. In: Dewick, P.M. (ed.) *Medicinal Natural Products: a Biosynthetic Approach*, 3rd edn. Wiley, Chichester, UK, pp. 137–186. doi: 10.1002/9780470742761

Dorsch, W., Muller, A., Christoffel, V., Stuppner, H., Antus, S., *et al.* (1994) Antiasthmatic acetophenones – an *in vivo* study on structure activity relationship. *Phytomedicine* 1, 47–54. doi: 10.1016/S0944-7113(11)80022-X

Feng Y., Cui, C., Liu, X., Wu Q., Hu, F., *et al.* (2017) Protective role of apocynin via suppression of neuronal autophagy and TLR4/NF-κB signaling pathway in a rat model of traumatic brain injury. *Neurochemical Research* 42(11), 3296–3309. doi: 10.1007/s11064-017-2372-z

Gardner, Z. and McGuffin, M. (eds) (2013) *Botanical Safety Handbook*. CRC Press, Boca Raton, Florida.

Garza, R. (2015) Creosote bush: medicinal shrub of the Southwest deserts. *Journal of the American Herbalists Guild* 13(1), 42–46.

Ghosh, S., Chisti, Y. and Banerjee, U.C. (2012) Production of shikimic acid. *Biotechnology Advances* 30, 1425–1431. https://doi.org/10.1016/j.biotechadv.2012.03.001

Greinwald, R. and Stobernack, H. (1990) *Ammi visnaga* (Khella). *British Journal of Phytotherapy* 1, 7–10.

Günaydin, K. and Beyazit, N. (2004) The chemical investigations on the ripe fruits of *Ammi visnaga* (Lam.) Lamarck, growing in Turkey. *Natural Product Research* 18(2), 169–175.

Habtemariam, S. (2018) Molecular pharmacology of rosmarinic and salvianolic acids: potential seeds for Alzheimer's and vascular dementia drugs. *International Journal of Molecular Sciences* 19(2), 458. doi: 10.3390/ijms19020458

Harborne, J. and Baxter, H. (1993) *Phytochemical Dictionary*. Taylor & Francis, London.

Hatcher, H., Planalp, R., Chob, J., Tortia, F.M. and Tortic, S.V. (2008) Curcumin: from ancient medicine to current clinical trials. *Cell Molecular Life Sciences* 65(11), 1631–1652. doi: 10.1007/s00018-008-7452-4

Hussain, H., Krohn, K., Ahmad, V.U., Miana, G.A. and Green, I.R. (2007) Lapachol: an overview. *Archive for Organic Chemistry* (ARKIVOC) (ii), 145–171. https://doi.org/10.3998/ark.5550190.0008.204

Jendželovská, Z., Jendželovský, R., Kuchárová, B. and Fedorocko, P. (2016) Hypericin in the light and in the dark: two sides of the same coin. *Frontiers in Plant Science* 7, 560. https://doi.org/10.3389/fpls.2016.00560

Khalil, N., Bishr, M., Desouky, S. and Salama, O. (2020) *Ammi Visnaga* L., a potential medicinal plant: a review. *Molecules* 25(2), 301. doi: 10.3390/molecules25020301

Kleemann, B., Loos, B., Scriba, T.J., Lang, D. and Davids, L.M. (2014) St. John's wort (*Hypericum perforatum* L.) photomedicine: hypericin-photodynamic therapy induces metastatic melanoma cell death. *PLoS ONE* 9(7): e103762. doi: 10.1371/journal.pone.0103762

Kubin, A., Wierrani, F., Burner, U., Alth, G. and Grünberger, W. (2005) Hypericin – the facts about a controversial agent. *Current Pharmaceutical Design* 11, 233–253. doi: 10.2174/1381612053382287

Kulkarni, S.S. and Cantó, C. (2015) The molecular targets of resveratrol. *Biochimica et Biophysica Acta* 1852(6), 1114–1123. doi: 10.1016/j.bbadis.2014.10.005

Lattanzioa, V., Kroonb, P.A., Linsalatac, V. and Cardinali, A. (2009) Globe artichoke: a functional food and source of nutraceutical ingredients. *Journal of Functional Foods* 1(2), 131–144. https://doi.org/10.1016/j.jff.2009.01.002

Liu, Y., Lund, J.A., Murch, S.J. and Brown, P.N. (2018) Single-lab validation for determination of kavalactones and flavokavains in *Piper methysticum* (kava). *Planta Medica* 84(16), 1213–1218. https://doi.org/10.1055/a-0637-2400

Lu, J.M., Nurko, J., Weakley, S.M., Jiang, J., Kougias, P., *et al.* (2010) Molecular mechanisms and clinical applications of nordihydroguaiaretic acid (NDGA) and its derivatives: an update. *Medical Science Monitor* 28 16(5), 93–100. Available at: https://www.ncbi.nlm.nih.gov/pmc/articles/PMC2927326/ (accessed 25 November 2020).

Lude, S., Torok, M., Dieterle, S., Jaggi, R., Buter, K.B. and Krahenbuhl, S. (2008) Hepatocellular toxicity of kava leaf and root extracts. *Phytomedicine* 15, 120–131. doi: 10.1016/j.phymed.2007.11.003

Luo, W., Ang, C.Y., Schmidt, T.C. and Betz, J.M. (1998) Determination of salicin and related compounds in botanical dietary supplements by liquid chromatography with fluorescence detection. *Journal of Obstetric Anaesthesia and Critical Care International* 81(4), 757–762. https://doi.org/10.1093/jaoac/81.4.757

Mahdi, J.G. (2010) Medicinal potential of willow: a chemical perspective of aspirin discovery. *Journal of the Saudi Chemical Society* 14(3), 317–322. https://doi.org/10.1016/j.jscs.2010.04.010

Mazieiro, R. Frizon, R.R. Barbalho, S.M. and Goulart, R.A. (2018) Is curcumin a possibility to treat inflammatory bowel diseases? *Journal of Medicinal Food* 21(11), 1077–1085. doi: 10.1089/jmf.2017.0146

Meier, B. and Liebi, M. (1990) Medicinal plants containing salicin: effectiveness and safety. *British Journal of Phytotherapy* 1, 36–42.

Mir, M.Y., Hamid, S., Kamili, A.N. and Hassan, Q.P. (2019) Sneak peek of *Hypericum perforatum* L.: phytochemistry, phytochemical efficacy and biotechnological interventions. *Journal of Plant Biochemistry and Biotechnology* 28, 357–373. https://doi.org/10.1007/s13562-019-00490-7

Nawaz, A., Khan, G.M., Hussain, A., Ahmad, A., Khan, A. and Safdar, M. (2011) Curcumin: a natural product of biological importance. *Gomal University Journal of Research* 27(1), 7–14. Available at: http://www.gomal.pk/GUJR/PDF/PDF-June-2011/2%20Nawaz%207-14%20Paper.pdf (accessed 25 November 2020).

Nelson, K.M., Dahlin, J.L., Bisson, J., Graham, J. and Pauli, G.F. (2017) The essential medicinal chemistry of curcumin. *Journal of Medicinal Chemistry* 60, 1620–1637. doi: 10.1021/acs.jmedchem.6b00975

Noleto-Dias, C., Ward, J.L., Bellisai, A., Lomax, C. and Beale, M.H. (2018) Salicin-7-sulfate: a new salicinoid from willow and implications for herbal medicine. *Fitoterapia* 127, 166–172. https://doi.org/10.1016/j.fitote.2018.02.009

Okamura, T., Okada, M., Kikuchi, T., Wakizaka, H. and Zhang, M.R. (2018) Kinetics and metabolism of apocynin in the mouse brain assessed with positron-emission tomography. *Phytomedicine* 38(1), 84–89. doi: 10.1016/j.phymed.2017.05.006

Ooi, S.L., Henderson, P. and Pak, S.C. (2018) Kava for generalized anxiety disorder: a review of current evidence. *Journal of Alternative and Complement Medicine* 24(8), 770–780. doi: 10.1089/acm.2018.0001

Panossian, A. and Wikman, G. (2008) Pharmacology of *Schisandra chinensis* Bail.: an overview of Russian research and uses in medicine. *Journal of Ethnopharmacology* 118(2), 183–212. doi: 10.1016/j.jep.2008.04.020

Park, E.J. and Pezzuto, J.M. (2015) The pharmacology of resveratrol in animals and humans. *Biochimica et Biophysica Acta* 1852(6), 1071–1113. https://doi.org/10.1016/j.bbadis.2015.01.014

Pluskal, T., Torrens-Spence, M.P., Fallon, T.R., De Abreu, A., Shi, C.H. and Weng, J.-K. (2019) The biosynthetic origin of psychoactive kavalactones in kava. *Nature Plants* 5, 867–878. https://doi.org/10.1038/s41477-019-0474-0

Rao, K.V., McBride, T.J. and Oleson, J.J. (1968) Recognition and evaluation of lapachol as an antitumour agent. *Cancer Research* 28(10), 1952–1954. Available at: https://cancerres.aacrjournals.org/content/28/10/1952.short (accessed 25 November 2020).

Reboredo-Rodríguez, P., Varela-López, A., Forbes-Hernández, T.Y., Gasparrini, M., Afrin, S., *et al.* (2018) Phenolic compounds isolated from olive oil as nutraceutical tools for the prevention and management of cancer and cardiovascular diseases. *International Journal of Molecular Sciences* 19, 2305. doi: 10.3390/ijms19082305

Ribnicky, D.M., Poulev, A. and Raskin, I. (2003) The determination of salicylates in *Gaultheria procumbens* for use as a natural aspirin alternative. *Journal of Nutraceuticals, Functional and Medicinal Foods* 4(1), 39–52. doi: 10.1300/j133v04n01_05

Ryu, B., Kim, H.M., Woo, J.H., Choi, J.H. and Jang, D.S. (2016) A new acetophenone glycoside from the flower buds of *Syzygium aromaticum* (cloves). *Fitoterapia* 115, 46–51. doi: 10.1016/j.fitote.2016.09.021

Saeed, Г., Mehjabeen, Jahan, M. and Ahmad, N. (2014) *In vivo* evaluation and safety profile evaluation of *Arctostaphylos uva-ursi* (L) Spreng extract in rabbits. *Pakistan Journal of Pharmaceutical Sciences* 27(6), 2197–2205. Available at: https://pdfs.semanticscholar.org/da51/4c0fcd73bb4f59c4dbc24144622b1dca5f44.pdf (accessed 25 November 2020).

Salem, B.M., Kolsi, R.B.A., Dhouibi, R., Ksouda, K., Charfi, S., *et al.* (2017) Protective effects of *Cynara scolymus* leaves extract on metabolic disorders and oxidative stress in alloxan-diabetic rats. *BMC Complementary and Alternative Medicine* 17(1), 328. doi: 10.1186/s12906-017-1835-8

Sarris, J., Kavanagh, D.J., Byrne, G., Bone, K.M., Adams, J. and Deed, G. (2009) The Kava Anxiety Depression Spectrum Study (KADSS): a randomized, placebo-controlled crossover trial using an aqueous extract of *Piper methysticum*. *Psychopharmacology* 205, 399–407. doi: 10.1007/s00213-009-1549-9

Siegers, C., Bodinet, C., Ali, S. and Siegers, C.P. (2003) Bacterial deconjugation of arbutin by *Escherichia coli*. *Phytomedicine* 10(Supplement 4), 58–60.

Teshcke, R. (2011) Kava and the risk of liver toxicity: past, current, and future. *American Herbal Products Association (AHPA) Report* 26(3), 9–17. Available at: https://www.docdroid.net/jHcw2HK/11-0303-march2011-ahpa-report-kava-special-report.pdf (accessed 25 November 2020).

Van Wyk, B.E. and Wink, M. (2017) *Medicinal Plants of the World*, 2nd edn. CAB International, Wallingford, UK.

Widhalm, J.R. and Rhodes, D. (2016) Biosynthesis and molecular actions of specialized 1,4-naphthoquinone natural products produced by horticultural plants. *Horticulture Research* 3, 16046. doi: 10.1038/hortres.2016.46

Willis, J.T. (2000) *Juglans* spp., juglone and allelopathy. *Allelopathy Journal* 7(l), l–55. Available at: http://www.allelopathyjournal.org/Journal_Articles/AJ%207%20(1)%20January,%202000%20(1–55).pdf (accessed 25 November 2020).

Xuan, T.D., Fukuta, M., Wei, A.C., Elzaawely, A.A., Khanh, T.D. and Tawata, S. (2008) Efficacy of extracting solvents to chemical components of kava (*Piper methysticum*) roots. *Journal of Natural Medicines* 62(2), 188–194. doi: 10.1007/s11418-007-0203-2

Polyphenols – Tannins and Flavonoids

3

Polyphenol compounds are those with two or more benzene rings, with varying degrees of hydroxylation in each ring. The name has become almost synonymous with dietary antioxidants. In Chapter 2, this volume, readers were introduced to lignans and stilbenes, both examples of polyphenol compounds. In this chapter, the focus is on the two largest subcategories of polyphenols – tannins and flavonoids.

Tannins

Tannins represent the largest group of polyphenols. They are widely distributed in the bark of trees, insect galls, leaves, stems and fruit. Tannins were originally isolated from the bark and insect galls of oak trees. They are the chief plant constituents responsible for astringency.

Tannins are non-crystalline compounds which in water produce a mild acid reaction. Their ingestion gives rise to a puckering, astringent sensation in the mouth and the taste is sour. Tannins are high-molecular-weight compounds (500–20,000 Da) containing sufficient phenolic hydroxyl groups to permit the formation of stable cross-links with proteins, and as a result of this cross-linking enzymes may be inhibited. Their ability to precipitate proteins into insoluble complexes enables humans to 'tan' animal hides and convert them to leather. This ability is also the basis of their astringent effects. Due to protein precipitation, the tannins exert an inhibitory effect on many enzymes, hence contributing a protective function in bark and heartwoods of woody plant species. Tannins also form precipitates with polysaccharides and some alkaloids, including caffeine.

Most tannins are readily extracted by water and ethanol. For research purposes microwave extraction has been shown to be quite effective, yielding high tannin levels for raspberry leaf (*Rubus idaeus*, Rosaceae) (Cobzac *et al.*, 2005).

DOI: 10.1079/9781789243079.0003

Tannins and astringency

Astringency has been defined as a loss of wettability of the thin mucous layer at the palate (Jobstl *et al.*, 2004). Astringency leads to contraction of tissues, binding and toning of skin and mucous membranes, drying up of exudations, formation of a thin protective layer when applied to wounds, precipitation of proteins and polysaccharides on the surface of membranes which harden the epidermis and reduce permeability.

The astringent action comes about when tannins form cross-linkages with salivary proteins and glycoproteins. The major proteins involved are proline-rich proteins (PRPs) and histatins (a family of histidine-rich proteins or HRPs), while α-amylase and mucins may also form complexes with tannins (Jobstl *et al.*, 2004; Bajec and Pickering, 2008). Histatins are of particular interest, since they are potent antimicrobials with an affinity for inhibition of *Candida albicans*, the fungus associated with oral thrush (Khurshid *et al.*, 2016). It is unclear how the formation of complexes with tannins modifies this activity.

In addition, tannins may interact with taste receptors, modifying the sensory qualities of tastes such as bitter and sweet (McRae and Kennedy, 2011).

Tannins as digestion inhibitors

The traditional view states that salivary proteins are induced in mammals by the presence of tannins, which render plants more difficult to digest. Salivary proteins protect digestive enzymes, an action that helps avoid nutritional deficiencies (Johns, 1990; Shimada, 2006). Since consumption of tannins may lead to reduced absorption of proteins and other nutrients, problems in compounding tannin-containing herbal medicines may occur. For this reason, it is not considered wise to drink tea with meals, particularly iced tea – an American favourite. Herb teas, on the other hand, are generally much lower in tannins compared to black tea (Blake, 1993/94). Some authors point to the numerous benefits of dietary tannins (Mueller-Harvey, 2006), and it is a well-established fact that humans have a long history of both enjoying tannin-rich foods and beverages, and consuming them to ease indigestion.

Classification of tannins

Historically, tannins have been classified as either hydrolysable tannins (HTs) or condensed tannins (CTs). In a more contemporary classification proposed by Khanbabaee and van Ree (2001), the hydrolysable category is replaced with the categories: **gallotannins** and **ellagitannins**. Complex tannins are a combination of two or more dissimilar tannin units.

HTs are derived from simple phenolic acids, namely **gallic** and **ellagic** acids. They contain at least two phenolic rings linked by esterification to one or more polyols (sugar alcohol derivatives with multiple hydroxyl groups). HTs are readily soluble in water or alcohol. Hydrolysed by water or by the action of acids or enzymes such as tannase, HTs give gallic acid and glucose or ellagic acid and glucose.

Gallotannins consist of galloyl units (formed from shikimic acid) linked to glucose residues (or other polyols).

Ellagitannins consist of ellagic acid units (formed from two or more galloyl units with depside (C–C) linkages. **Rugosidin** and **oenothein B** are examples of ellagitannins.

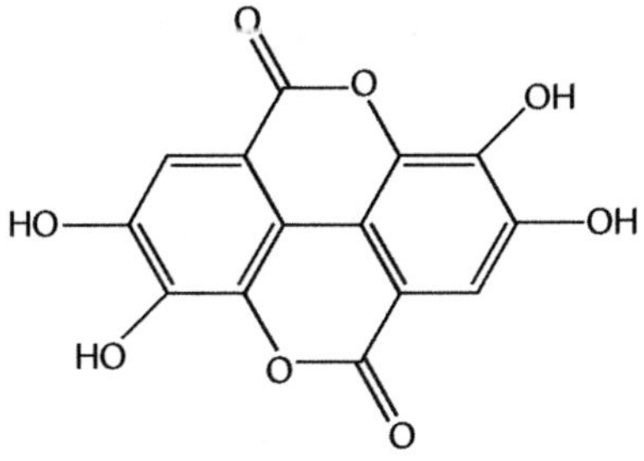

ellagic acid

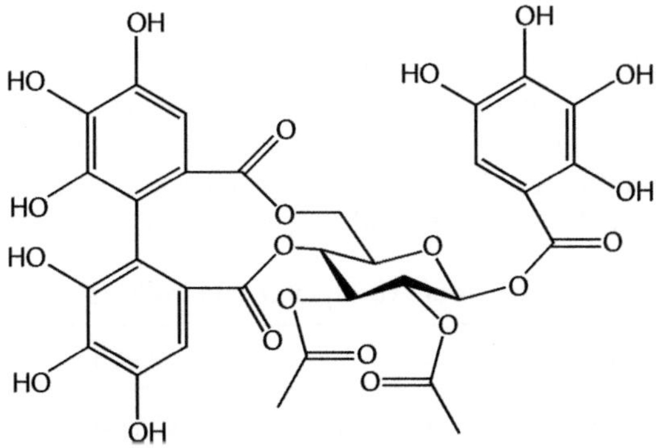

rugosidin D – section of molecule

Actions of HTs include:

- broad spectrum antimicrobial activity;
- protection of inflamed mucous membranes;
- drying effect on mucous membranes, reduces hypersecretions;
- reduce inflammation and swelling which accompany infections;
- prevent bleeding from small wounds;
- reduce uterine bleeding, for example in menorrhagia, metrorrhagia;
- binding effect in the gut – relieves diarrhoea, dysentery; and
- external cleansing – used as douches, snuffs, eyewash, throat gargle.

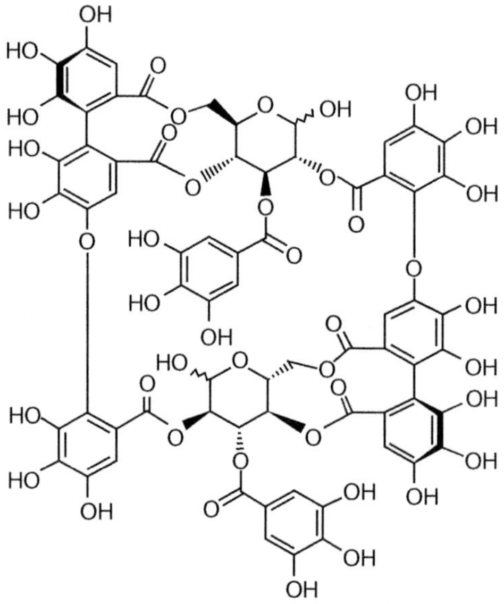

oenothein B – an ellagitannin

Apart from the general actions listed above, some specific actions for a range of HTs are listed in Table 3.1.

Condensed tannins (CTs) or **phlobotannins** are polymers of flavanols (catechins) and flavan-3,4-diols (leucoanthocyanins). The monomeric catechins and leucoanthocyanidins have insufficient hydroxyl groups to create the cross-linkages necessary for precipitating proteins, hence they are not technically tannins but flavonoids. However, their oligomers and polymers do have tanning properties (Khanbabaee and van Ree, 2001). Upon hydrolysis CTs

Table 3.1. Some specific actions of hydrolysable tannins (HTs)

Compound	Plant origin	Action
Sanguiin	*Rubus idaeus* (Rosaceae)	Cytotoxic, antidiarrhoeic
Rugosin D	*Filipendula ulmaria* (Rosaceae)	Antitumour
Gallotannic acid	*Quercus* spp. (Fagacaeae)	Astringent, antihaemorrhagic and antimicrobial
Oenthein B	*Epilobium* spp. (Onagraceae), *Eucalyptus* spp. (Myrtaceae)	Inhibitor of 5α-reductase and aromatase, anti-inflammatory, antitumour
Hamamelitannin	*Hamamelis virginiana* (Hamamelidaceae)	Antioxidant, anti-inflammatory
Geraniin	*Geranium maculatum* (Geraniaceae), *Phyllanthus* spp. (Phyllanthaceae)	Cytotoxic, hypolipidaemic
Agrimoniin	*Agrimonia eupatoria* (Rosaceae)	Astringent, antidiarrhoeic

form insoluble red residues or phlobaphenes. They are only partially soluble in water and alcohol, while the addition of glycerine aids solubility.

Oligomeric procyanidins (OPCs) are a major subcategory of CTs in which two or more molecules of catechin and/or epicatechin are linked by simple carbon bonds. OPCs are present in popular beverages such as green and black teas and red wine. They are responsible for many of the benefits associated with green tea, including the promotion of oral health and findings of lower mortality from some forms of cerebrovascular disease, cardiovascular disease (CVD) and cancer (Cabrera *et al.*, 2006). Green tea is especially rich in the beneficial epigallocatechin (EGC), which is a potent antioxidant and vascular tonic.

OPCs are extracted commercially from grape seeds and pine bark. The high-profile product under the proprietary name of **Pycnogenol®**, was first isolated from *Pinus pinaster* (Pinaceae) by Professor Jacques Masquelier of the University of Bordeaux, while searching for agents that might enhance or extend the activity of vitamin C (Masquelier *et al.*, 1979). Since then, numerous controlled clinical trials of Pycnogenol®, which is standardized to contain over 70% OPCs, have demonstrated improvement of endothelial function in coronary artery disease, platelet function normalization for chronic venous insufficiency, hypertension and its complications as well as numerous non-cardiovascular disorders (Oliff, 2019).

Grape skins, flesh and seed are all sources of characteristic tannins in wine. Those from the skins have the highest molecular weights, including up to 83 flavanol subunits, composed of procyanidins and prodelphinidins – the latter consists mainly of EGC with trace levels of gallocatechin and epigallocatechin 3-O-gallate (McRae and Kennedy, 2011).

Wine polyphenols have been widely investigated with respect to human health, though it should be noted that other non-tannin polyphenols may also contribute to such effects, including flavonoids such as quercetin, the stilbene resveratrol, and hydroxytyrosol as discussed in Chapter 2, this volume. Wine in moderation is cardioprotective, due in part to the significant enhancement of eNOS (endothelial nitric oxide synthase) expression and nitric oxide (NO) release by polyphenols, since reduced NO is linked to onset of atherosclerosis, a major cause of heart disease (Leikert *et al.*, 2002).

Apart from tea and wine, one of the mostly widely studied of the polyphenol-rich plants is the pomegranate (*Punica granatum*, Lythraceae). While the leaves, bark, roots and flowers have been used in medicine, the fruit provides three polyphenol medicinal products: the seed, juice and pericarp (or peel) (Lansky and Newman, 2007). In a recent study, 79 phenolic compounds from pomegranate peel were described, the most numerous of which were ellagitannins, along with gallotannins, OPCs including procyanidin and prodelphinidin dimers, anthocyanins, flavonoids and phenolic acids (Ambigaipalan *et al.*, 2016). The most potent phenolic compounds in pomegranate peel are the complex ellagitannins, such as punicalagin (2,3-HHDP-4,6-gallagylglucoside) containing glucose, ellagic acid and gallagic acid.

Other ellagitannins identified include punicafolin and punigluconin (Lansky and Newman, 2007; Fisher *et al.*, 2011; Ambigaipalan *et al.*, 2016).

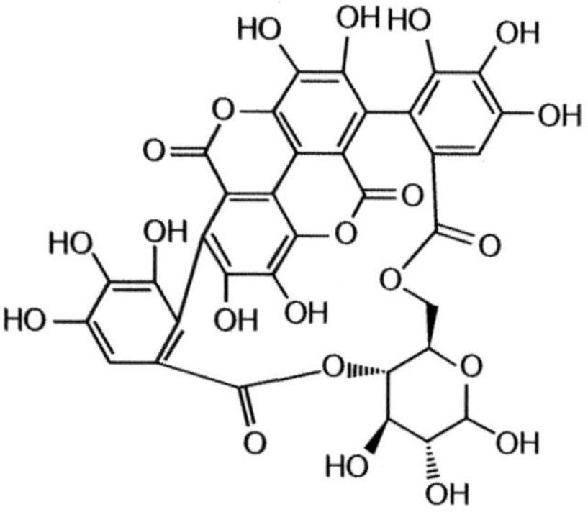

punicalagin – an ellagitannin

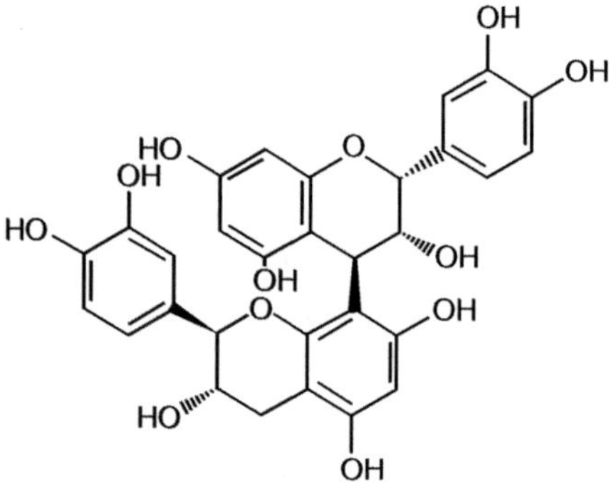

procyanidin B1 – an OPC

Pomegranate polyphenols have been studied for cancer prevention and treatment and anti-inflammatory effects (Lansky and Newman, 2007). **Punicalagin**, responsible for over 50% of the antioxidant properties of pomegranate juice, inhibits growth of human oral, prostate and colon tumour cells, however, not as effectively as the pure juice, indicating the likelihood of synergistic effects between the polyphenols and other constituents (Seeram *et al.*, 2005).

Along with grapes and pomegranate, almost all edible fruit species with dark (red, purple, blue) colours have been shown to be rich in polyphenols, organic acids and vitamins, and many are widely promoted as dietary antioxidants and even 'superfoods'. In this context, the Australian Kakadu plum (*Terminalia ferdinandiana*, Combretaceae) is a stand-out. Kakadu plum not only contains the highest known levels of vitamin C in the plant kingdom, it also contains very high levels of phenolic compounds, more than six times the level of blueberry (Netzel *et al.*, 2007; Konczak *et al.*, 2010). Blueberry is considered the standard reference for both the total phenolic content (TPC) and corresponding antioxidant capacity (Konczak, 2017). Kakadu plum's rich mixture of polyphenols include ellagitannins such as corilagin and puncialin, along with flavonoids such as **quercetin** and **luteolin**. Polyphenols from Kakadu plum leaves are potent inhibitors of bacterial species known to trigger autoimmune reactions, applicable to diseases such as rheumatoid arthritis, ankylosing spondylitis and multiple sclerosis (Courtney *et al.*, 2015).

Antibacterial properties

Tannins are known to be potent antimicrobials. Different mechanisms of action enable them to inhibit biofilm formation and quorum sensing in pathogenic bacteria. Tannins also inhibit the activity of β-lactamase, the enzyme responsible for multidrug resistance against many standard antibiotics (Mandal *et al.*, 2010). The most potent of the polyphenols tested were tannic acid, epicatechin and epigallocatechin gallate (EGCG).

Flavonoids

As noted in Chapter 2, this volume, flavonoids are products of both the shikimic acid pathway and the acetate pathway, being formed by the condensation of a phenylpropanoid precursor with three malonyl coenzyme A units. Their chemical structure is based on a C_{15} skeleton consisting of two benzene rings connected by a three-carbon chain, that is, C_6–C_3–C_6. The three-carbon chain is generally closed to form a heterocyclic ring (the C-ring), and flavonoid categories are classified on the basis of degree of oxidation and other structural variations within the C-ring (see Table 3.2). Note this table includes flavonols such as catechins, already reviewed in the tannins section above. Plants typically contain a series of closely interrelated flavonoids, with different degrees of hydroxylation and methyloxylation substitution in the A- and B-rings.

Flavonoids occur both in the free state and as glycosides, for example **rutin** is the glycoside of quercetin, the aglycone (non-sugar portion). They are yellow and white plant pigments (Latin *flavus* = yellow). Rutin was discovered in rue (*Ruta graveolens*, Rutaceae) in 1842 – it later became known as vitamin P (permeability factor).

Role in plant physiology

Flavonoids are universal within the plant kingdom – they are the most common plant pigments next to chlorophyll and carotenoids. They are recognized as the pigments responsible for autumnal leaf colours as well as for the many shades of yellow, orange and red in flowers. Their functions include protection of plant

Table 3.2. Classification of flavonoids based on C (middle)-ring structure (from Beecher, 2003; Bhagwat et al., 2015)

Subclass	Structural feature in C-ring[a]	Examples	Structure	Plant sources
Flavone	Oxo at C-4; double bond at C-2–C-3	Apigenin Luteolin Baicalein	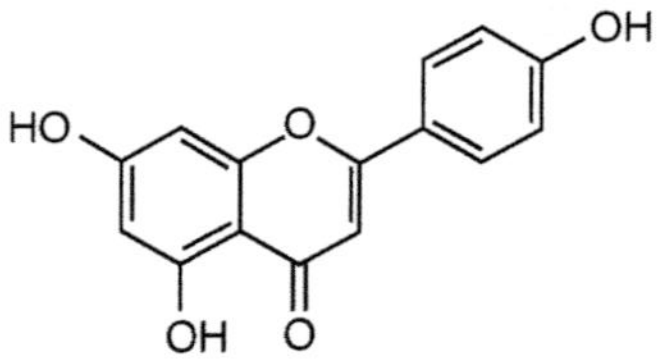 flavone	Chamomile Parsley Agrimony Passionflower Skullcap
Flavanone	Oxo at C-4; no double bonds	Naringenin Eriodictyol Hesperetin Liquiritin	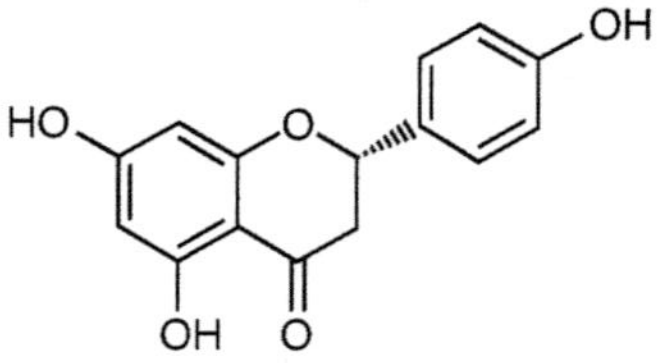 flavanone	Citrus Liquorice
Flavonol	Hydroxyl at C-3; oxo at C-4; double bond at C-2–C-3	Quercetin Kaempferol Myricetin Isorhamnetin	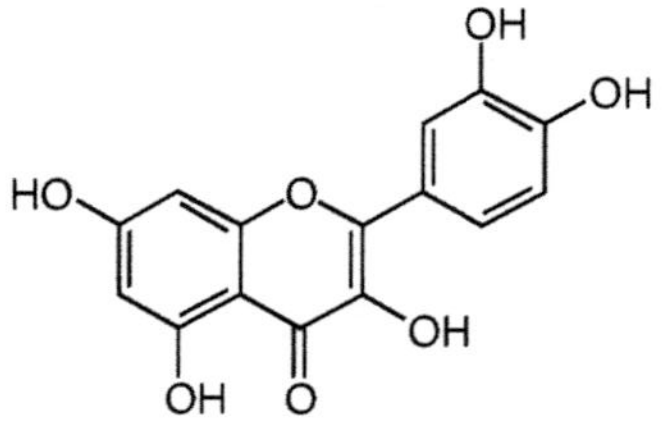 flavonol	St John's wort Hawthorn Oregano Dill Celery

			Tea Red grapes Red wine

Flavarol | Hydroxyl at C-3; no double bonds | Catechin
Epicatechin |

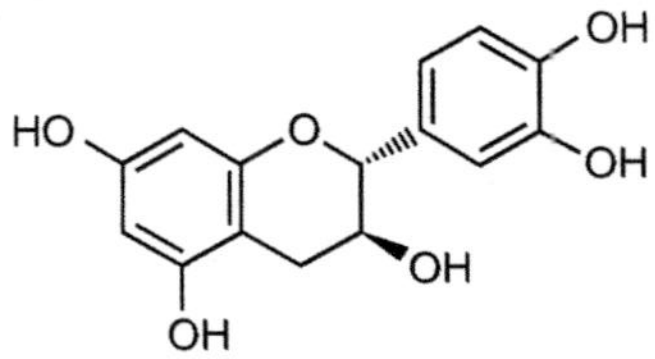

flavanol

			Green tea Cocoa beans Hawthorn St John's wort

Flavan-3-ol (OPCs) | O-gallate at C-3; no double bond | Epicatechin-3-gallate
Epigallocatechin-3-gallate |

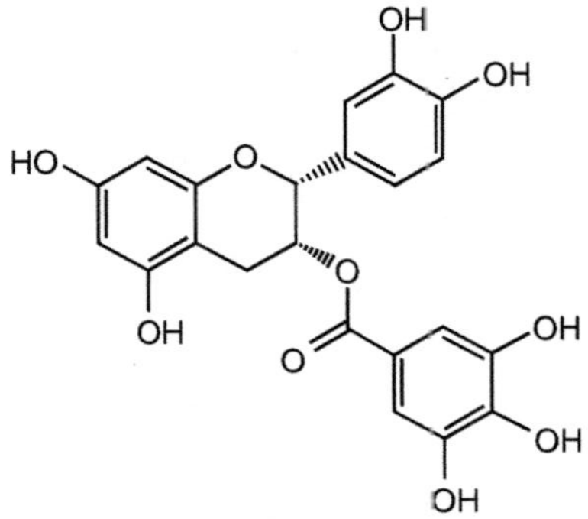

flavan-3-ol

Continued

Table 3.2. Continued.

Subclass	Structural feature in C-ring[a]	Examples	Structure	Plant sources
Anthocyanidins	Hydroxyl at C-3; double bonds at C-1–C-2 and C-3–C4	Malvadin Cyanidin	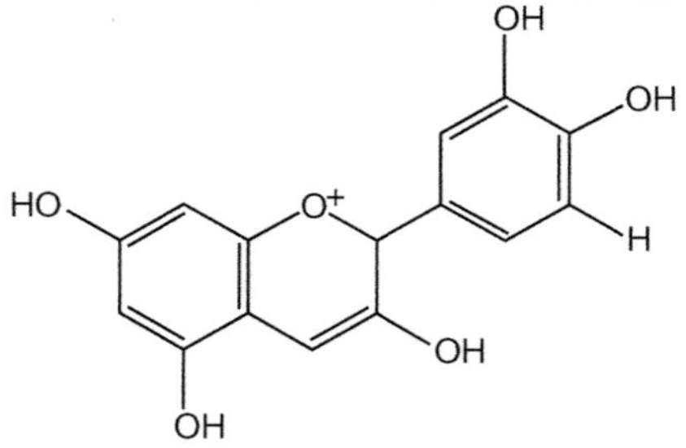 cyanidin	Bilberry Most red, blue and purple flowers and fruit
Isoflavone	C-ring: oxo at C-4; double bond at C-2–C-3; B-ring linked at C-3	Genistein Formononetin Diadzein	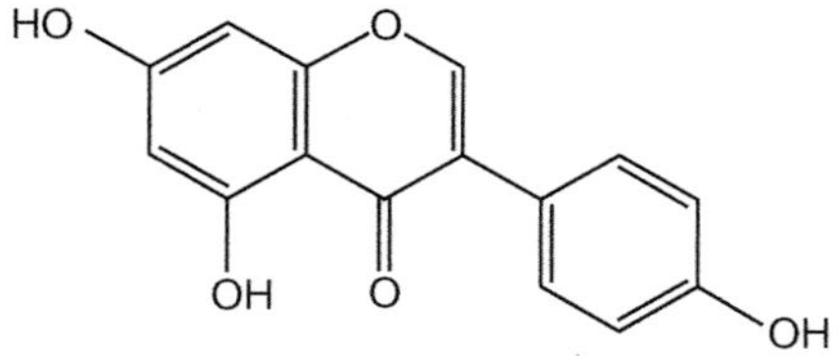 genistein – an isoflavone	Soybean Liquorice Red clover

OPCs, oligomeric procyanidins.

[a]oxo, A prefix in the formal International Union of Pure and Applied Chemistry (IUPAC) nomenclature for the functional group '=O' (a substituent oxygen atom connected to another atom by a double bond).

tissues from damaging UV radiation, acting as antioxidants, enzyme inhibitors, pigments and light screens. The compounds are involved in photosensitization and energy transfer, action of plant growth hormones and growth regulators, as well as defence against infection (Middleton, 1988). The plant response to injury results in increased synthesis of flavonoid aglycones (including phyto-alexins) at the site of injury or infection.

Flavonoids and the human diet

Flavonoids can be considered as important constituents of the human diet. There are numerous databases in existence that quantify the levels of flavon-oids in various foods and beverages, and for estimates of average daily intakes from the diet. The United States Department of Agriculture (USDA) holds one of the biggest databases (Bhagwat *et al.*, 2015). Unfortunately, for various reasons the daily intake levels between databases is rather inconsistent, how-ever, the mean intake of flavonoids for the USA was calculated at 251 mg daily, most intake coming from tea in the form of flavan-3-ols (Sebastian *et al.*, 2015).

Therapeutics

Experiments have proven flavonoids affect the heart and circulatory system and strengthen the capillaries. They are often referred to as 'biological stress modi-fiers' since they serve as protection against environmental stress (Middleton, 1988). They are also known to have synergistic effects with ascorbic acid. Their protective actions are due in part to membrane stabilizing and antioxi-dant effects.

Therapeutic effects of flavonoids such as antioxidant, antiviral, hepato-protective, anti-atheromatous, anti-inflammatory and antihypertensive have been widely reported, though these effects are dependent on their degree of absorption. Rutin and hespiridin among others are effective in reducing permeability of blood capillaries and are widely used for peripheral vascular disorders.

In Holland, the Zutphen Elderly Study (Hertog *et al.*, 1993) was carried out on 805 men aged 65–84 years. Dietary intake of flavonoids was calculated during a 5-year period. The study indicates that the intake of flavonoids is inversely associated with mortality from coronary heart disease and, to a lesser extent, myocardial infarct. The main sources of flavonoids in the men's diets were tea, onions and apples. These flavonoids include **quercetin, kaempferol** and **myricitin** as well as flavonols (CTs) found in black tea (Hertog *et al.*, 1993). The findings have been reproduced in subsequent studies, a meta-analysis found evidence for an inverse relationship for risk of CVD with the six main structural classes of flavonoids (Wang *et al.*, 2014).

Epidemiological studies with dietary flavonoids have also demonstrated an inverse association with incidence of stroke (Rice-Evans, 2001). Antitumour activity has been demonstrated in flavonoids. **Nomilin** in citrus induces glutathione-S-transferase (GST), which aids in detoxification of carcinogens. Quercetin inhibits cytotoxic T lymphocytes *in vitro*. *In vitro* studies also suggest flavonoids such as tangeretin and nobiletin from citrus fruits may inhibit mutagenesis produced by mutagens from cooked food, thereby playing a role in prevention of carcinogenesis (Calomme *et al.*, 1996).

Animal studies show many flavonoids protect liver cells from damage induced by toxins such as carbon tetrachloride. Silymarin, the flavonolignan complex from *Silybum marianum* (Asteraceae), is an inducer of Phase 1 detoxication, and protects liver mitochondria and microsomes from lipid peroxidation. This protection also occurs with the flavonoids quercetin and taxifolin. Silymarin causes hepatocyte regeneration and increases hepatic glutathione *in vivo*, while clinical studies demonstrated benefits in subjects with alcoholic liver diseases, cirrhosis and viral hepatitis (Pradhan and Girish, 2006).

Flavonoids of all categories have been found to act as neuroprotective agents. They show great promise for treatment and prevention of neurodegenerative disorders such as Parkinson's disease (Jung and Kim, 2018).

Enzyme inhibitors

The presence of numerous hydroxyl functional groups on their two benzene rings enables flavonoids to easily attach to amino acid residues on enzyme surfaces, leading to potent inhibition of some enzyme systems in humans (Miller, 1973; Scotti *et al.*, 2012). Examples include:

1. Aldose reductase – causes diabetic cataracts. Quercetrin is regarded as a very potent inhibitor of this enzyme. Others include nepetrin and its glycoside nepitin from *Rosmarinus officinalis* (Lamiaceae), as well as glycosides of luteolin and kaempferol (Pathak *et al.*, 1991).

2. Xanthine oxidase – causes hyperuricaemia leading to gout and renal stones. It is postulated that free hydroxyl groups at C-5 and C-7 are important for inhibition of xanthine oxidase (Pathak *et al.*, 1991).

3. Tyrosine protein kinase and other protein kinases (PKs) – these are involved in the division, proliferation, metabolism and apoptosis within cells, while PK inhibitors such as quercetin and kaempferol are involved in the downregulation of various signalling molecules and pathways that lead to the initiation of various forms of cancer (Scotti *et al.*, 2012). The isoflavone genistein is another reputed tyrosine kinase inhibitor, however, only at unrealistically high doses (Cooke *et al.*, 2006).

4. Lipoxygenase and cyclooxygenase – involved with production of inflammatory prostaglandins, leukotrienes and thromboxanes, derivatives of arachidonic acid.

5. Cyclic nucleotide phosphodiesterase (PDE) – a key enzyme in promotion of platelet aggregation.

6. Phase I (cytochrome P450) and Phase II metabolizing enzymes – these enzymes are involved in the metabolism of xenobiotics and carcinogens, which can lead to their inactivation. While this mechanism contributes to the cancer-protective properties of flavonoids, it may also have implications for drug inter-actions, including during chemotherapy (Moon *et al.*, 2006).

Quercetin is an effective inhibitor of histamine release, when induced by various agents. It inhibits different stages of inflammation including granula-tion tissue formation in chronic arthritis (Middleton and Drzewiecki, 1984). Flavonoids offer the advantage of a high margin of safety, and lack of side effects such as ulcerogenicity over the classical anti-inflammatory drugs.

Diseases associated with increased permeability of blood capillaries include diabetes, chronic venous insufficiency, scurvy, haemorrhoids, varicose ulcers and bruising.

Buckwheat (*Fagopyrum esculentum*, Polygonaceae) contains 8% rutin if grown under suitable ecological conditions. Tissue examinations of animals with induced oedema plaques in corneal and conjunctival tissue showed that after treatment with 180 mg rutin from buckwheat herb (tablets) the oedema was flushed out and the fragility of the vascular system was reduced (Schilcher and Muller, 1982).

Lipid peroxidation, the oxidative degradation of polyunsaturated fatty acids, is implicated in several pathological conditions – hepatotoxicity, haemolysis, cancer, atherosclerosis, tumour promotion and inflammation. Selected flavon-oids, along with the flavonolignan silymarin, may exert protective effects against cell damage produced by lipid peroxidation stimulated by a variety of toxins, owing to the antioxidant properties of the compounds (Pathak *et al.*, 1991).

Isoflavones

Isoflavones are flavonoid isomers whose distribution is largely restricted to the Fabaceae (legume) family. They have structural similarities to oestrogens so they are also classed among the phytoestrogens, compounds that bind to oestrogen receptors, but whose oestrogen activity is relatively low.

Clinical trials have supported the efficacy of isoflavones, especially **genis-tein** found in soymilk and other soy products, in prevention and treatment of breast, prostate and other cancers (Sarkar *et al.*, 2006) as well as alleviating hot flushes and other symptoms associated with the menopause (Albert *et al.*, 2002). Other investigations have focused on the interactions between isoflavones and other flavonoids on signalling pathways involving oestrogen receptors (ERα, ERβ), nuclear receptors PPARs and estrogen receptor-related receptors (ERRs) all of which can influence the amount of oestrogen response within particular cells, and potentially provide a model for the variable effects that have been observed on the endocrine and immune systems (Cooke *et al.*, 2006).

Anthocyanins

Anthocyanin pigments are found in flowers and red, blue and black fruits. They are present in the plant as glycosides of hydroxylated 2-phenylbenzopyrylium or flavylium salts. The aglycones (non-sugar portion) are known as anthocyanidins – only six are of widespread occurrence. Chemical structure has a significant influence on anthocyanin colours. As the number of sugar hydroxyl and methyl groups in the B-ring (at the top right) increases, the colour changes from orange to blue as the maximum visible absorption shifts to longer wavelengths. Hence, **cyanidin** turns orange-red and **delphinidin** bluish-red in methanolic solutions (Mazza and Miniati, 1993). The presence of flavone co-pigments, chelating metals or aromatic acyl substitutes tends to produce a blueing effect (Cseke and Kaufman, 1999).

Rich sources of anthocyanins are grape skins and bilberry (*Vaccinium myrtillus*, Ericaceae) – along with other members of the *Vaccinium* genus, including blueberries. Bilberry has long been associated with effects on microcirculation and is used in diabetic neuropathy and ophthalmology. Despite the presence of OPCs and other flavonoids, experimental and clinical studies have found most of the potency of bilberry lies in the anthocyanin fraction. A superoxide anion scavenging effect has been demonstrated *in vitro* and *in vivo* for bilberry anthocyanins (Martín-Aragón *et al.*, 1999), while numerous anthocyanin-containing berries were shown to possess superoxide radicals scavenging and antilipoperoxidant activities (Constantino *et al.*, 1994). In an *in vivo* study, anthocyanins from black rice were found to promote atherosclerotic plaque and collagen-stabilizing activity, and improving lipid profiles (Xia *et al.*, 2006).

Intake of berries in general, anthocyanidins and total flavonoids in an elderly population is associated with reduction in cognitive decline (Devore *et al.*, 2012).

Bioavailability of polyphenols

Polyphenols are, with some exceptions, poorly absorbed in humans, meaning that only a fraction of the quantity ingested actually reaches the circulation, and the tissues or cells that are the site of intended action (D'Archivio *et al.*, 2010). One consequence of this is the interpretation of results based on *in vitro* studies, since the polyphenols being tested are not subject to metabolism, as they are for *in vivo* studies. To further complicate matters, we often have a paucity of information regarding rates of absorption and metabolism, the nature of the metabolites and their potential biological activities (Lila, 2004; Williams *et al.*, 2004).

Flavonoids often occur in the form of glycosides (i.e. linked to sugars), which are generally poorly absorbed until undergoing hydrolysis by bacterial enzymes in the large intestine, where they are subject to ring fission by intestinal bacteria, a process in which the middle (non-aromatic) carbon ring is split apart into smaller fission metabolites. These metabolites are readily absorbed, and some are known to possess therapeutic benefits in their own right (Bone,

1995). For example the metabolites of OPCs from grape seed were studied in humans, and found to be predominantly phenolic acids (Ward *et al.*, 2004). Other flavonoids such as quercetin are metabolized more rapidly in the small intestine, either by enzymatic action or by glucose transporters (Day *et al.*, 2000; Gee and Johnson, 2001).

Other factors influencing the bioavailability of polyphenols include: (i) environmental factors such as degree of ripeness of fruit; (ii) processing factors such as cooking; (iii) the food matrix whereby other food or herbal ingredients may enhance or inhibit the absorption; (iv) the structure and amount of the polyphenol ingested; and (v) host-related factors such as the microflora profile, gender, age, pre-existing illness and so on (D'Archivio *et al.*, 2010). Finally, it should be remembered that flavonoids do not occur in isolation; rather they are in a phytochemical matrix whether they are found in plant foods, herbs or beverages. Synergistic and potentiating actions between flavonoids and other phytochemicals may not only enhance (or inhibit) their absorption and bioavailability, but also enhance their biological and therapeutic activities (Lila, 2004).

References

Albert, A., Altabre, C., Baro, F., Buendia, E., Cabero, A., *et al.* (2002) Efficacy and safety of a phytoestrogen preparation derived from *Glycine max* (L.) in climacteric symptomatology. *Phytomedicine* 9(2), 85–92. doi: 10.1078/0944-7113-00107

Ambigaipalan, P., de Camargo, F. and Shahidi, F. (2016) Phenolic compounds of pomegranate byproducts (outer skin, mesocarp, divider membrane) and their antioxidant activities. *Journal of Agriculture and Food Chemistry* 64(34), 6584–6604. doi: 10.1021/acs.jafc.6b02950

Bajec, M.R. and Pickering, G.J. (2008) Astringency: mechanisms and perception. *Critical Reviews in Food Science and Nutrition* 48(9), 1–18. doi: 10.1080/10408390701724223

Beecher, G.R. (2003) Overview of dietary flavonoids: nomenclature, occurrence and intake. In: *Proceedings of the Third International Scientific Symposium on Tea and Human Health: Role of Flavonoids in the Diet. Journal of Nutrition* 133(10), 3248S–3254S. doi: 0.1093/jn/133.10.3248S

Bhagwat, S., Haytowitz, D.B. and Holden, J.M. (2015) *USDA Database for the Flavonoid Content of Selected Foods.* Release 3.2 (November 2015). United States Department of Agriculture (USDA), Beltsville, Maryland. Available at: https://data.nal.usda.gov/dataset/usda-database-flavonoid-content-selected-foods-release-32-november-2015 (accessed 28 November 2020).

Blake, O. (1993/94) The tannin content of herbal teas. *British Journal of Phytotherapy* 3, 124–127.

Bone, K. (1995) Oestrogen modulation. *The Modern Phytotherapist* 1, 8–10.

Cabrera, C., Reyes, R. and Giménez, R. (2006) Beneficial effects of green tea – a review. *Journal of the American College of Nutrition* 25(2), 79–99. doi: 10.1080/07315724.2006.10719518

Calomme, M., Pieters, L., Vlietinck, A. and Vanden Berghe, D. (1996) Inhibition of bacterial mutagenesis by *Citrus* flavonoids. *Planta Medica* 62(3), 222–226. doi: 10.1055/s-2006-957864

Cobzac, S., Moldovan, M., Olah, N.K., Bobos, L. and Surducan, E. (2005) Tannin extraction efficiency, from *Rubus idaeus, Cydonia oblonga* and *Rumex acetosa,* using different extraction techniques and spectrophotometric quantification. *Acta Universitatis Cibiniensis Seria F Chemia* 8(2), 55–59.

Constantino, L., Rastelli, G., Rossi, T., Bertoldi, M. and Albasini, A. (1994) Composition, superoxide radicals scavenging and antilipoperoxidant activity of some edible fruits. *Fitoterapia* LXV, 44–47.

Cooke, P.S., Selvaraj, V. and Yellayi, S. (2006) Genistein, estrogen receptors, and the acquired immune response. *The Journal of Nutrition* 136(3), 704–708. https://doi.org/10.1093/jn/136.3.704

Courtney, R., Siraarta, J., Matthews, B. and Cook, I.E. (2015) Tannin components and inhibitory activity of kakadu plum leaf extracts against microbial triggers of autoimmune inflammatory diseases. *Pharmacognosy Journal* 7(1), 18–31. doi: 10.5530/pj.2015.7.2

Cseke, L.J. and Kaufman, P.B. (1999) Regulation of metabolite synthesis by plants. In: Kaufman, P.B., Cseke, L.J., Warber, S., Duke, J.A. and Brielmann, H.L. (eds) *Natural Products from Plants.* CRC Press, Boca Raton, Florida, pp. 101–141.

D'Archivio, M., Filesi, C., Varì, R., Scazzocchio, V. and Masella, R. (2010) Bioavailability of the polyphenols: status and controversies. *International Journal of Molecular Sciences* 11(4), 1321–1342. doi: 10.3390/ijms11041321

Day, A.J., Cañada, F.J., Díaz, J.C., Kroon, P.A., Mclauchlan, R., *et al.* (2000) Dietary flavonoid and isoflavone glycosides are hydrolysed by the lactase site of lactase phlorizin hydrolase. *FEBS Letters* 468(2–3), 166–170. doi: 10.1016/s0014-5793(00)01211-4

Devore, E.E., Kang, J.H., Breteler, M.M.B. and Grodstein, F.G. (2012) Dietary intake of berries and flavonoids in relation to cognitive decline. *Annals of Neurology* 72(1), 135–143. doi: 10.1002/ana.23594.

Fisher, U.A., Carle, R. and Kammerer, D.R. (2011) Identification and quantification of phenolic compounds from pomegranate (*Punica granatum* L.) peel, mesocarp, aril and differently produced juices by HPLC-DAD–ESI/MS^n. *Food Chemistry* 127(2), 807–821. https://doi.org/10.1016/j.foodchem.2010.12.156

Gee, J.M. and Johnson, I.T. (2001) Polyphenolic compounds. Interactions with the gut and implications for human health. *Current Medicinal Chemistry* 8(11), 1245–1255. doi: 10.2174/0929867013372256

Hertog, M.G., Feskens, J.M., Hollman, P.C.H., Katan, M.B. and Krombout, D. (1993) Dietary antioxidant flavonoids and risk of coronary heart disease. *The Lancet* 342(8878), 1007–1011. doi: 10.1016/0140-6736(93)92876-u

Jobstl, E., O'Connell, J., Patrick, J., Fairclough, A. and Williamson, M.P. (2004) Molecular model for astringency produced by polyphenol/protein interactions. *Biomacromolecules* 5, 942–949.

Johns, T. (1990) *The Origins of Human Diet and Medicine.* University of Arizona Press, Tucson, Arizona.

Jung, U.J. and Kim, S.R. (2018) Beneficial effects of flavonoids against Parkinson's disease. *Journal of Medicinal Food* 21(5), 421–432. doi: 10.1089/jmf.2017.4078

Khanbabaee, K. and van Ree, T. (2001) Tannins: classification and definition. *Natural Products Reports* 18(6), 641–649. doi: 10.1039/b101061l

Khurshid, Z., Naseem, M., Sheikh, Z., Najeeb, S., Shahab, S. and Zafar, M.S. (2016) Oral antimicrobial peptides: types and role in the oral cavity. *Saudi Pharmaceutical Journal* 24(5), 515–524. doi: 10.1016/j.jsps.2015.02.015

Konczak, I. (2017) Health attributes of indigenous Australian plants. In: Cherikoff, V. (ed.) *Wild Foods. Looking Back 60,000 Years for Clues to Our Future Survival.* New Holland Publications, London, pp. 130–155. Available at: http://www. wildfoodscience.com/ (accessed 28 November 2020).

Konczak, I., Zabaras, D., Dunstan, M. and Aguas, P. (2010) Antioxidant capacity and hydrophilic phytochemicals in commercially grown native Australian fruits. *Food Chemistry* 123(4), 1048–1054. https://doi.org/10.1016/j.foodchem.2010.05.060

Lansky, E.P. and Newman, R.E. (2007) *Punica granatum* (pomegranate) and its potential for prevention and treatment of inflammation and cancer. *Journal of Ethnopharmacology* 109, 177–206. doi: 10.1016/j.jep.2006.09.006

Leikert, J.S., Räthel, T.R., Wohlfart, P., Cheynier,V., Vollmar, A.M. and Dirsch, V.M. (2002) Red wine polyphenols enhance endothelial nitric oxide synthase expression and subsequent nitric oxide release from endothelial cells. *Circulation* 106, 1614–1617. doi: 10.1161/01.CIR.0000034445.31543.43

Lila, M.A. (2004) Anthocyanins and human health: an *in vitro* investigative approach. *Journal of Biomedicine and Biotechnology* 2004(5), 306–313. doi: 10.1155/S111072430440401X

Mandal, S.M., Dias, R.O. and Franco, O.L. (2010) Phenolic compounds in antimicrobial therapy. *Journal of Medicinal Food* 20(10), 1031–1038. https://doi.org/10.1089/jmf.2017.0017

Martín-Aragón, S., Basabe, B., Benedí, J.M. and Villar, A.M. (1999) *In vitro* and *in vivo* antioxidant properties of *Vaccinium myrtillus*. *Pharmaceutical Biology* 37(2), 109–113. https://doi.org/10.1076/phbi.37.2.109.6091

Masquelier, J., Michaud, J., Laparra, J. and Dumon, M.C. (1979) Flavonoids and pycnogenols. *International Journal of Vitamin and Nutrition Research* 49(3), 307–311.

Mazza, G. and Miniati, E. (1993) *Anthocyanins in Fruits, Vegetables and Grains*. CRC Press, Boca Raton, Florida. doi: 10.3390/molecules16042348

McRae, J.M. and Kennedy, J.A. (2011) Wine and grape tannin interactions with salivary proteins and their impact on astringency: a review of current research. *Molecules* 16(3), 2348–2364. doi: 10.3390/molecules16032348

Middleton Jr, E. (1988) Plant flavonoid effects on mammalian cell systems. In: Cracker, F.E. and Simon, F.E. (eds) *Herbs, Spices and Medicinal Plants*, Vol. 3. Oryx Press, Phoenix, Arizona, pp. 103–144.

Middleton, E. and Drzewiecki, G. (1984) Flavonoid inhibition of human basophil histamine release stimulated by various agents. *Biochemical Pharmacology* 33(21), 3333–3338. doi: 10.1016/0006-2952(84)90102-3

Miller, F.P. (ed.) (1973) *Phytochemistry*, Vol. 2. Van Nostrand Reinhold Co., New York.

Moon, Y., Wang, X. and Morris, M.E. (2006) Dietary flavonoids: effects on xenobiotic and carcinogen metabolism. *Toxicology in Vitro* 20, 187–210.

Mueller-Harvey, I. (2006) Unravelling the conundrum of tannins in animal nutrition and health. *Journal of the Science of Food and Agriculture* 86(13), 2010–2037. https://doi.org/10.1002/jsfa.2577

Netzel, M., Netzel, G., Tian, Q., Schwartz, S. and Konczak, I. (2007) Sources of antioxidant activity in Australian native fruits. Identification and quantification of anthocyanins. *Journal of Agricultural and Food Chemistry* 54(26), 9820–9826. doi: 10.1021/jf0622735

Oliff, H. (2019) *Scientific and Clinical Monograph for Proprietary Botanical Ingredient Pycnogenol®*. American Botanical Council, Austin, Texas.

Pathak, D., Pathak, K. and Gingla, A.K. (1991) Flavonoids as medicinal agents – recent advances. *Fitoterapia LXII*, 371–389.

Pradhan, S.C. and Girish, C. (2006) Hepatoprotective herbal drug, silymarin from experimental pharmacology to clinical medicine. *Indian Journal of Medical Research* 124(5), 491–504.

Rice-Evans, C. (2001) Flavonoid antioxidants. *Current Medicinal Chemistry* 8(7), 797–807. doi: 10.2174/0929867013373011

Sarkar, F.H., Adsule, S., Padhye, S., Kulkarni, S. and Li, Y. (2006) The role of genistein and synthetic derivatives of isoflavone in cancer prevention and therapy. *Mini Reviews in Medicinal Chemistry* 6(4), 401–407. doi: 10.2174/138955706776361439

Schilcher, H. and Muller, A. (1982) *Fagopyrum esculentum* Moench – a new pharmaceutically useful flavonoid plant. *Studies in Organic Chemistry* 11, 523–528.

Scotti, L., Mendonça Jr, F.J.B., Moreira, D.R.M., da Silva, M.S., Pitta, I.R. and Scotti, M.T. (2012) SAR, QSAR and docking of anticancer flavonoids and variants: a review. *Current Topics in Medicinal Chemistry* 12(24), 1–25. doi: 10.2174/1568026611212240007

Sebastian, R.S., Enns, C.W., Goldman, J.D., Martin, C.L., Steinfeldt, L.C., Murayi, T. and Moshfegh, A.J. (2015) New database facilitates characterization of flavonoid intake, sources, and positive associations with diet quality among US adults. *The Journal of Nutrition* 145(6), 1239–1248. doi: 10.3945/jn.115.213025.

Seeram, N.P., Adams, L.S., Henning, S.M., Niu, Y., Zhang, Y., *et al.* (2005) *In vitro* anti-proliferative, apoptotic and antioxidant activities of punicalagin, ellagic acid and a total pomegranate tannin extract are enhanced in combination with other polyphenols as found in pomegranate juice. *The Journal of Nutritional Biochemistry* 16(6), 360–367. https://doi.org/10.1016/j.jnutbio.2005.01.006

Shimada, T. (2006) Salivary proteins as a defense against dietary tannin. *Journal of Chemical Ecology* 32(6), 1149–1163. doi: 10.1007/s10886-006-9077-0

Wang, X., Ouyang, Y.Y., Liu, J. and Zhao, G. (2014) Flavonoid intake and risk of CVD: a systematic review and meta-analysis of prospective cohort studies. *British Journal of Nutrition* 111(1), 1–11. doi: 10.1017/S000711451300278X

Ward, N.C., Croft, K.D., Puddey, I.B. and Hodgson, J.M. (2004) Supplementation with grape seed polyphenols results in increased urinary excretion of 3-hydroxyphenylpropionic acid, an important metabolite of proanthocyanidins in humans. *Journal of Agricultural Food Chemistry* 52(17), 5545–5549. https://doi.org/10.1021/jf049404r

Williams, R.J., Spencer, J.P.E. and Rice-Evans, C. (2004) Flavonoids: antioxidants or signalling molecules? (Part of a series of reviews on: 'Flavonoids and isoflavones (phytoestrogens): absorption, metabolism, and bioactivity'). *Free Radical Biology and Medicine* 36(7), 838–849. http://dx.doi.org/10.1016/j.freeradbiomed.2004.01.001

Xia, X., Ling, W., Ma, J., Xia, M., Hou, M., *et al.* (2006) An anthocyanin-rich extract from black rice. *Journal of Nutrition* 21(5), 421–432.

Glycosides

Introduction

Glycosides are a group of compounds consisting of a sugar portion (or moiety) attached by a special bond to one or more non-sugar portions. Chemically, they are hydroxyls of a sugar capable of forming ethers with other alcohols, or esters with acids.

Glycosides are broken down upon hydrolysis with enzymes or acids to:

1. a sugar moiety = glycone; and
2. a non-sugar moiety = aglycone/active portion.

These may be phenol, alcohol or sulfur compounds.

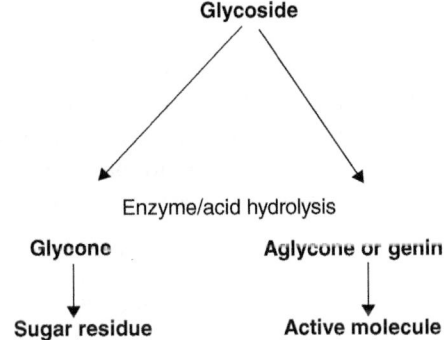

The bond between the two moieties may involve a phenolic hydroxyl group in which case an O-glycoside is formed. In other cases carbon (C-glycosides), nitrogen (N-glycosides) or sulfur (S-glycosides) may be involved. Glycosidic bonds are of extra significance since they link monosaccharides together to form oligosaccharides and polysaccharides.

Most glycosides can be classed as 'prodrugs' since they remain inactive until they are hydrolysed in the large bowel with the help of specialized bacteria, leading to the release of the aglycone – the truly active constituent.

DOI: 10.1079/9781789243079.0004

Classification of glycosides is based on the nature of the aglycone, which can be any of a wide range of molecular types, including phenols, quinones, terpenes and steroids. Since glycosides are so heterogeneous in structure they are not easy to learn as a specific group and are described together here for convenience. Most of the aglycone examples given here are also referred to in the chapters relating to their structural types.

Distribution

Glycoside distribution is widespread throughout the plant kingdom. They occur in the seeds of pulses, in swollen underground roots or shoots (yams, sweet potatoes), flowers and leaves. Some are toxic, especially cyanogenic and cardiac glycosides. Cooking may render them non-toxic. Glycosides are mostly soluble in water and organic solvents, though the aglycones tend to be somewhat insoluble in water.

Despite the widespread distribution of some glycoside classes we often find that the same botanical families consistently contain the same aglycone types, for example:

- Brassicaceae – glucosinolates;
- Rosaceae – cyanogenic glycosides;
- Scrophulariaceae – phenylpropanoid, iridoid, cardiac glycosides;
- Asteraceae – phenylpropanoid, flavonoid glycosides; and
- Polygonaceae – anthraquinone glycosides.

Cyanogenic glycosides

In cyanogenic glycosides the element nitrogen occurs in the form of nitriles (as a prefix, cyano- means nitrile), functional groups in which nitrogen is bound to carbon with triple bonds. Structurally they are derived from amino acids. The aromatic amino acid L-phenylalanine is the precursor to **amygdalin**, which retains the aromatic ring structure, whereas the non-aromatic amino acid, valine, is the precursor to **linamarin**, a cyanogesic glycoside with no aromatic ring.

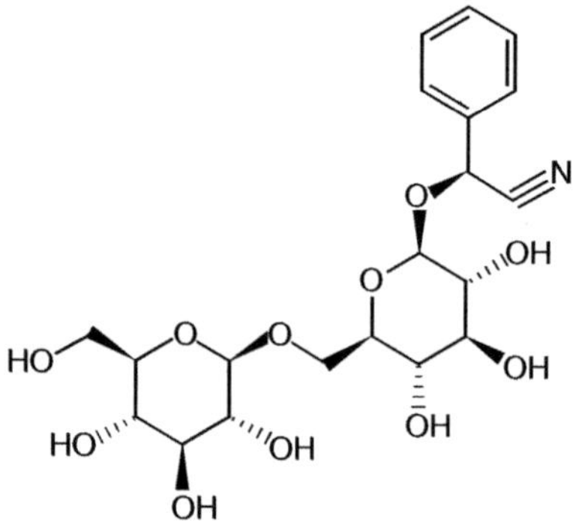

amygdalin

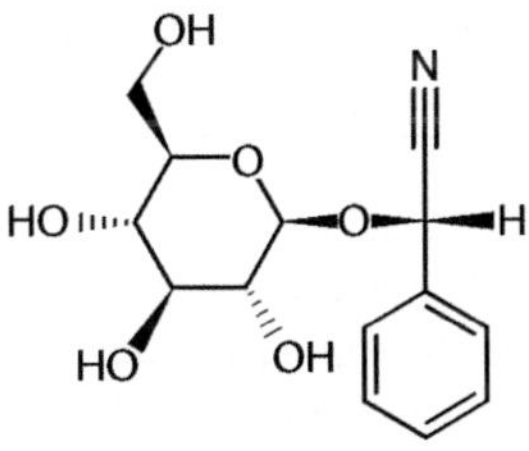

prunasin

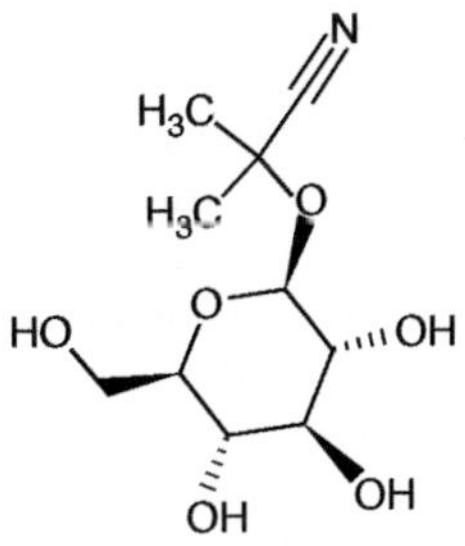

linamarin

Ingestion or processing of plants containing these glycosides releases hydrolytic β-glucosidase enzymes, leading to the formation of toxic hydrogen cyanide (HCN). Amygdalin, for example, is hydrolysed in the presence of the enzyme amygdalase and water, involving a two-stage process to produce glucose, along with an aglycone made up of benzaldehyde (scent of bitter almonds) and odourless HCN.

The occurrence of cyanogenic glycosides is widespread. Amygdalin and **prunasin** are very common among plants of the Rosaceae, particularly the genus *Prunus*, including bitter almond (*Prunus dulcis* var. *amara*) and wild cherry bark (*Prunus serotina*). Also included are the kernels of apricots, peaches and plums. Marzipan flavour is derived from amygdalin. Linamarin is found in linseed (*Linum usitatissimum*, Linaceae) and cassava (*Manihot esculenta*, Euphorbiaceae), a traditional flour in South America which is also widely cultivated in South-east Asia. The glycosides are characteristic of several other plant families including the Poaceae (grasses) and Fabaceae (legumes) (Table 4.1).

Sambunigrin (D-mandelonitrile glucoside) is found in leaves of the elder tree (*Sambucus nigra*, Viburnaceae) – it is isomeric to prunasin.

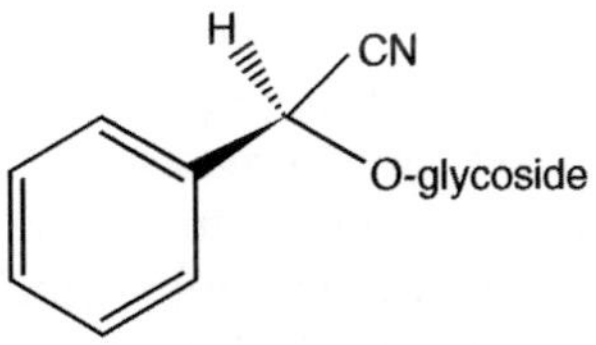

sambunigrin

Table 4.1. Major food sources of cyanogenic glycosides (from Dewick, 2009; Bolarinwa *et al.*, 2016)

Constituent	Plant/food source	Precursor amino acid
Amygdalin, prunasin	Bitter almonds (*Prunus dulcis* var. *amara*), seeds of other *Prunus* spp. (apricot, peach, cherry), apple seeds (*Malus* spp.) and wild cherry bark (*Prunus serotina*)	L-phenylananine
Sambunigrin	Leaves of elder tree (*Sambucus nigra*, Viburnaceae)	L-phenylananine
Linamarin	Linseed (*Linum usitatissimum*, Linaceae) and cassava (*Manihot esculenta*, Euphorbiaceae)	L-valine
Lotaustralin	Cassava (*M. esculenta*, Euphorbiaceae)	L-isoleucine
Dhurrin	Sorghum (*Sorghum bicolor*, Poaceae)	L-tyrosine
Taxiphyllin	Bamboo shoots (*Phyllostachys* spp., Poaceae)	L-tyrosine
Heterodendrin	Wattle seed (*Acacia* spp., Fabaceae)	L-leucine

Therapeutic actions

An amygdalin-containing drug called laetrile has been used as a cytotoxic agent in cancer, though its use is now restricted in most countries. A systematic review found no evidence that laetrile was beneficial to cancer patients (Milazzo *et al.*, 2007), however, it continues to be used in some circles.

In small quantities cyanic glycosides exhibit expectorant, sedative and digestive properties. Wild cherry bark, from *P. serotina* is an excellent cough remedy and tonic, as well as a flavouring agent used in cough syrups. It is of benefit as a tea for bronchitis. The main antitussive principle is prunasin, however, its levels are reduced in the process of drying the bark (Gardner and McGuffin, 2013).

Toxicology

HCN toxicity involves inactivation of the respiratory enzymes, leading initially to dizziness and facial flushing. As few as five chewed bitter almonds or apricot kernels may prove fatal to a child, rather more for adults (Wink and van Wyk, 2008). Our bodies can neutralize cyanides by converting them to thiocyanates, which are eliminated in the urine (Bruneton, 1995), however, this capacity can be overloaded if doses of cyanide are sufficiently high. In very high doses the whole of the central nervous system ceases to function, and death follows.

Humans from different regions and cultures have learned to detoxify these compounds using a variety of processing methods, including cooking, grinding, soaking, fermenting and drying (Bolarinwa *et al.*, 2016). Detoxification requires the enzyme rhodanese plus sulfur-containing amino acids, for HCN to be converted to thiocyanate. Problems may arise from ingestion of improperly processed food sources, and may be amplified in poorly nourished individuals, who are deficient in sulfur amino acids, and therefore unable to efficiently detoxify the cyanide.

Phenylpropanoid glycosides

Classic phenylpropanoid glycosides (PhGs) have only been known since 1964 when the first – **verbascoside** – was isolated from *Verbascum sinuatum* (Scrophulariaceae). Since then over 200 PhGs have been reported, but verbascoside is the most widespread, having been identified in more than 60 species from 14 plant families. These glycosides – sometimes referred to as phenylethanoids – consist of three basic units:

- a central glucose;
- a C_6C_2 moiety, usually a dihydroxy-2-ethanol; and
- a C_6C_3 moiety, usually a hydroxycinnamic acid.

The aromatic units can be differentially derived and other saccharides are usually linked to one or two of the free hydroxyls of the central glucose. The precursors to the non-sugar moieties are tyrosine and cinnamic acid – products of the shikimic acid pathway (Cometa *et al.*, 1993).

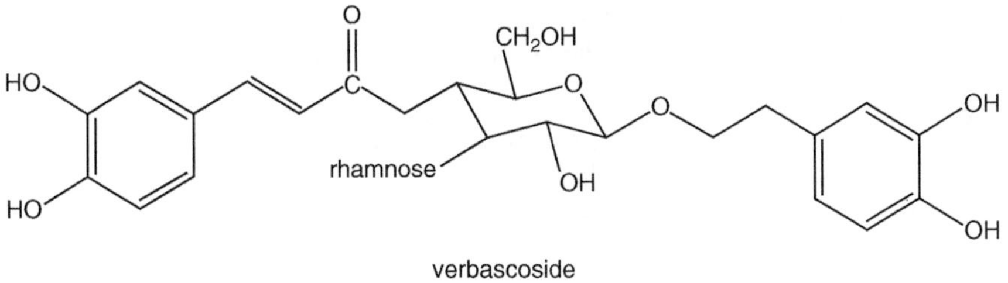

verbascoside

PhGs are common in the following families:

- Scrophulariaceae – *Verbascum* spp.;
- Orabanchaceae – *Rehmannia glutinosa*;
- Plantaginaceae – *Plantago asiatica, Plantago lanceolata; Digitalis purpurea*;
- Asteraceae – *Echinacea pallida, Echinacea angustifolia*;
- Lamiaceae – *Stachys* spp., *Teucrium* spp., *Ocimum sanctum*;
- Verbenaceae – *Verbena* spp., *Lantana camara*; and
- Oleaceae – *Syringia vulgaris, Forsythia* spp.

Leaves and flowers of several species of *Verbascum* are used therapeutically, including the common weed, *Verbascum thapsus*. For flower preparations, other species with larger flowers are preferred, including *Verbascum phlomoides* and *Verbascum densiflorum*. Levels of verbascoside in flowers vary between species, being quite low in *V. phlomoides* (Klimek *et al.*, 2010).

Therapeutic actions

Eight PhGs from steamed *Rehmannia glutinosa* demonstrated immunosuppressive activity *in vivo* – the most potent of these was verbascoside (Sasaki *et al.*, 1989).

In other studies, verbascoside demonstrated mild to moderate antitumour activity, as well as analgesic and neurosedative actions (Pieretti *et al.*, 1992). The antitumour activity is linked to inhibition of PKC (protein kinase C), an enzyme involved in cellular proliferation and differentiation (Herbert *et al.*, 1991). Verbascoside and other PhGs have shown activity against many cancer cell types, often with a biphasic effect – cytostatic and/or cytotoxic – depending on cell type (Pan *et al.*, 2003).

Echinocoside, a trisaccharide (similar to verbascoside with an extra sugar) found in *Echinacea angustifolia* and *E. pallida*, has proven antibiotic and antiviral properties. Echinocoside and other phenylpropane derivatives of *Echinacea* are potent antioxidants – together they protect skin against collagen degradation as a result of UV damage (Facino *et al.*, 1995).

Numerous other activities have been reported for PhGs, particularly from *Plantago* spp. and *Forsythia* spp. A comprehensive review of the pharmacological activities of this group is available (Cometa *et al.*, 1993).

Given the simple structures and high reactivity of phenylpropane molecules, a variety of glycoside forms exist – including many that don't conform to the classic PhG structure. **Syringin** – also known as eleuthroside B – from *Eleutherococcus senticosus* (Araliaceae) is an example. In this case glucose is part of a functional group attached to the benzene ring.

syringin – eleuthroside B

Several phenylpropanoid derivatives have been identified as contributing to the adaptogenic activity (adaptability to stress) in species such as *E. senticosus*, *Rhodiola rosea* (Crassulaceae) and *Ocimum sanctum* (Lamiaceae) (Wagner *et al.*, 1994). These non-specific but well-documented effects have been linked to the structural similarity of phenylpropanoids to catecholamines such as epinephrine and L-dopa. Ongoing investigations into *R. rosea* by Panossian and co-workers have demonstrated neuro-cardio- and hepato-protective activity for the phenylethanoid glycosides **rhodioloside** and **salirodoside** (Panossian *et al.*, 1999, 2010). In a more recent review, evidence was presented for pharmacological effects of salirodoside and *R. rosea* extract for age-related disorders, including Alzheimer's disease, Huntington's disease, depression, cardiovascular disease and diabetes (Zhuang *et al.*, 2019).

Anthraquinones

Anthraquinones are yellow-brown pigments, most commonly occurring as O-glycosides or C-glycosides. Their aglycones (anthrones) consist of two or more benzene rings linked by a quinone ring and marked by a pair of carbonyl functional groups in the C-9 and C-10 positions. Hydroxyl groups always occur at positions 1- and 8-, making them 1,8-dihydroxyanthraquinones. They have a wide distribution within the plant kingdom (Table 4.2), also occurring in fungi and lichens. Anthraquinone-containing plants have been used as dyes for textiles, for example the dyer's madder (*Rubia tinctorum*, Rubiaceae). They are also known as anthracene glycosides, since **anthracene** was the first compound isolated, by French chemists Dumas and Lambert, in 1832 (Clar, 1964).

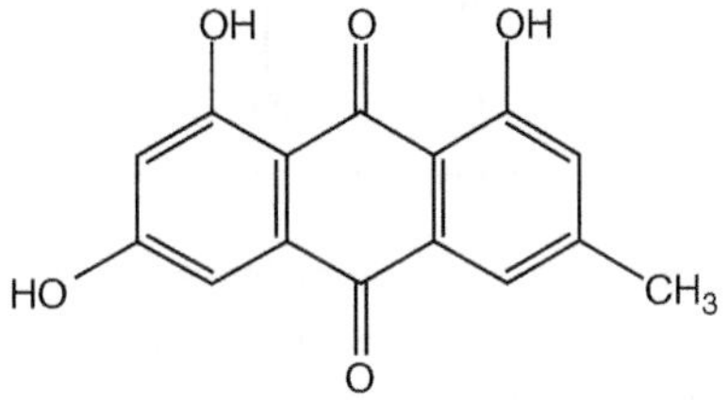

rheum-emodin

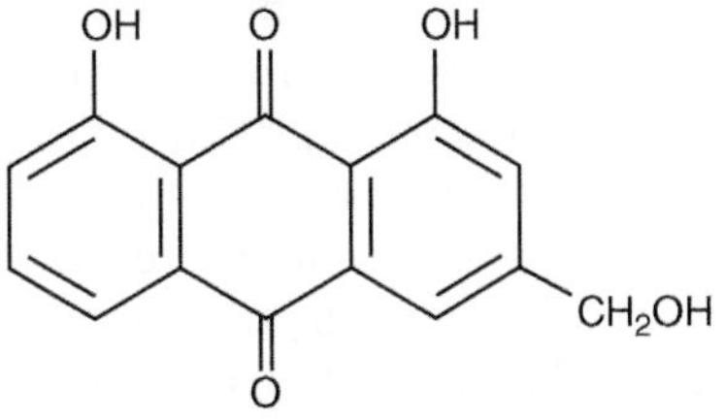

aloe-emodin

In dianthrones, the quinone function is lost through oxidative coupling of two anthrone molecules to form rhein dianthrones aka **sennidins A and B**, a process that only occurs after a plant is harvested and dried (Dewick, 2009). O-glycosylation of sennidins produces **sennosides**, the most potent cathartic compounds in *Senna*, usually present at 3–4%.

Table 4.2. Medicinal plant sources of anthraquinone glycosides

Species	Family	Anthraquinones
Frangula purshiana	Rhamnaceae	Cascarosides A–F, aloe-emodin
Rhamnus cathartica	Rhamnaceae	Glucofranulans A–D, dianthrones
Senna alexandrina	Fabaceae	Sennosides A–F, aloe-emodin
Rheum palmatum	Polygonaceae	Rheum-emodin, psyscion, chrysophanol
Aloe vera	Asphodelaceae	Aloin, aloe-emodin, barbaloin
Rumex crispus	Polygonaceae	Emodin, chrysophanol

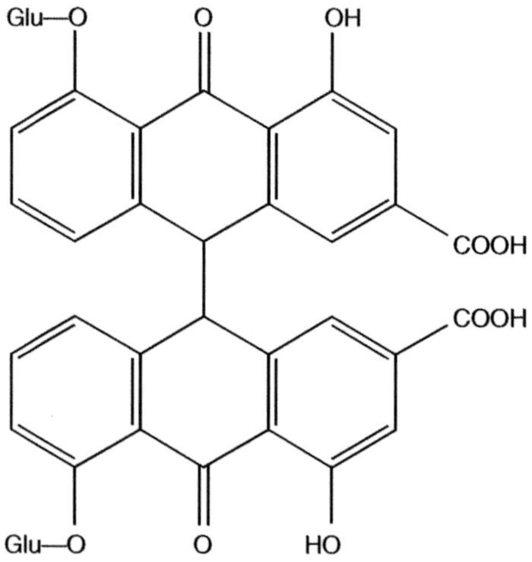

sennoside A, B

Apart from dianthrone glycosides, senna leaves and pods contain monomeric anthrone glycosides and their aglycones, such as **rhein-** and **aloe-emodin**.

Experimental investigations with the most widely prescribed anthraquinones – sennosides A and B – show they pass through the stomach and small intestine unaltered, but that in the caecum and colon they are converted to dianthrones by microorganisms. The dianthrones, which remain unabsorbed, are further transformed into anthrones, producing hydragogue and laxative effects in the process (Adzet and Camarasa, 1988).

The laxative effect is thought to occur as a result of increased peristaltic action and inhibition of water and electrolyte resorption by the intestinal mucosa. There is no evidence of direct irritation of the bowel mucosa (Bruneton, 1995).

Therapeutic actions

Anthraquinones are noted for their cathartic (laxative) effects. While plant sources tend to contain mixtures of anthraquinone glycosides and their aglycones. Aglycones have anti-inflammatory actions, but not the cathartic effect found in their glycosides.

The composition of glycosides and their derivatives in anthraquinone-containing plants determines their effectiveness as laxatives. The gentlest acting laxatives in this group belong to the buckthorns (*Rhamnus cathartica* and *Frangula frangula*, Rhamnaceae) and rhubarb (*Rheum palmatum*, Polygonaceae). In both cases the herbs are aged for at least a year during which the more irritant anthraquinone derivatives are converted to milder acting compounds. The presence of tannins also tends to moderate the laxative effect. The antiseptic effects of anthraquinones deter the growth of enteric pathogens.

Rumex crispus (Polygonaceae) or yellow dock is another rather mild cathartic, it contains anthrones such as **emodin**, as well as 1,5-dihydroxyanthraquinones, in which the second hydroxyl is in the C-5 position rather than C-8 (Günaydin *et al.*, 2002). Anthraquinones from *R. crispus* have shown promise as sun protection agents. Emodin, being the most potent, had an SPF (sun protection factor) rating of 30.59 at a concentration of 200 µg/ml (Demirezer and Uzun, 2016).

Aloes (*Aloe vera*, Asphodelaceae) and *Senna* spp. (Fabaceae) are the other commonly used laxative agents in this class. Senna leaves and pods are indicated for emptying bowels before X-ray investigations, and before and after abdominal operations (Bissett, 1994). Senna syrup is commonly prescribed for children and may be used during pregnancy and lactation for limited periods. Otherwise anthraquinones are contraindicated during pregnancy. The duration of action is around 8 hours, and they are usually taken before bed at night.

Safety

Due to the stimulant effect of these laxatives, they are contraindicated in irritable/spastic colon conditions. A slight overdose can produce griping and discomfort, an effect that may be counterbalanced by the presence of carminatives such as peppermint or coriander oil. It is unwise to rely on these remedies alone when treating chronic constipation, since dependence can result. Anthraquinone-containing preparations are contraindicated for patients with eating disorders. Chronic use in these circumstances can potentially result in kidney failure and death (Gardner and McGuffin, 2013).

Glucosinolates (mustard oil glycosides)

These are pungent-tasting compounds found mainly in the Brassicaceae family. They are sulfur-containing compounds synthesized from amino acids, and they have been classified according to the amino acid precursor (Cartier and Valesco, 2008):

- aromatic – tyrosine or phenylalanine;
- indole – tryptophan; and
- aliphatic – methionine.

Gluconasturtiin, **glucobrassicin** and **glucoraphanin** are examples of aromatic, indole and aliphatic glucosinolates, respectively. These and other glucosinolates and their precursors are shown in Table 4.3.

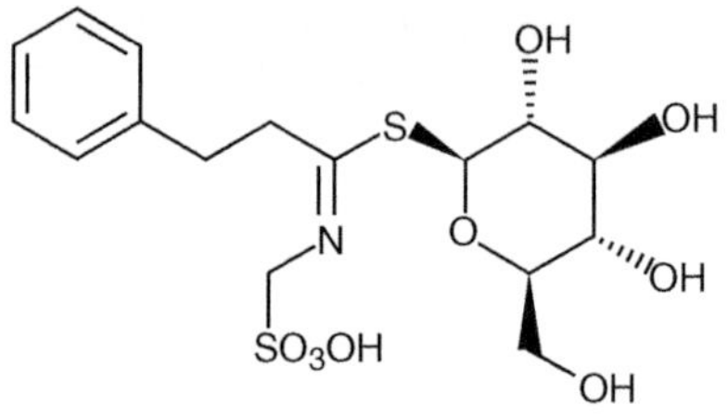

gluconasturtiin – aromatic

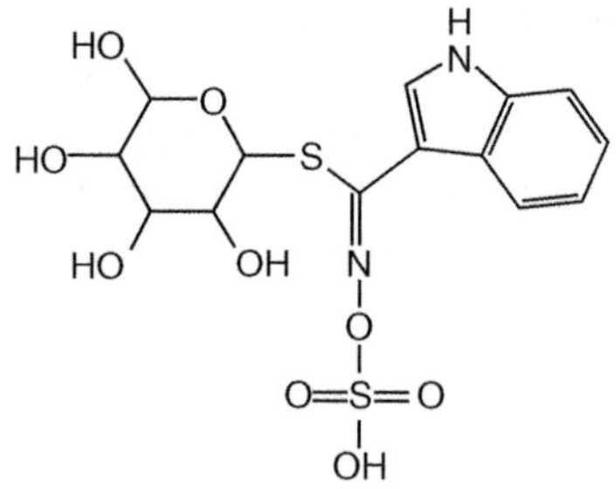

glucobrassicin – indole

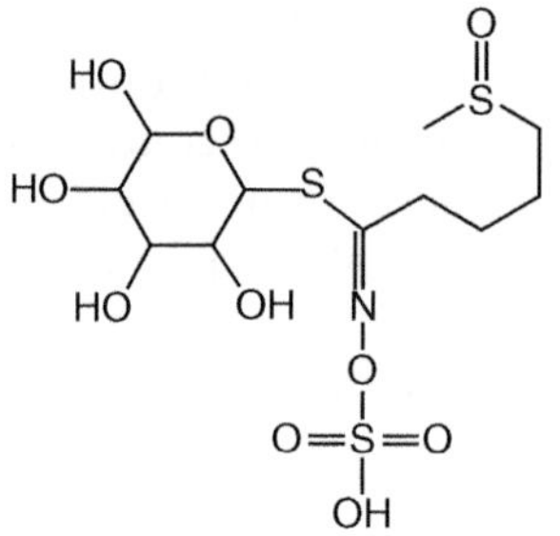

glucoraphanin – aliphatic

Sinigrin (potassium isothiocyanate), the glycoside from seeds of the black mustard (*Brassica nigra*, Brassicaceae), is hydrolysed on bruising or heating by the enzyme myrosinase to an unstable aglycone – **allyl isothiocyanate**. Depending on conditions other thiocyanates and highly toxic nitriles may be formed, the latter when plants are subjected to very hot water (> 45°C). A huge variability in relative abundance of the glycosides and their degradation products exists, associated with factors such as pH, season and genetic variety (Macleod, 1976; Cartier and Valesco, 2008).

At least 300 species of brassicas have been studied for their glucosinolate content. The compounds are mainly concentrated in the seeds (e.g. **sinalbin** from the seeds of white mustard (*Sinapis alba*, Brassicaceae)), although they can be found anywhere in the plants. They can always be identified by their spicy, pungent taste – responsible for the flavour of mustard seeds, horseradish root, rocket leaves as well as cabbage and its relatives such as broccoli. Aromatic glucosinolates occur in the garden nasturtium (*Tropaeolum majus*, Tropaeolaceae) and cress (*Lepidium sativum*, Brassicaceae), in the form of **glucotropaeolin**, which is hydrolysed to the antibiotic compound **benzyl isothiocyanate**.

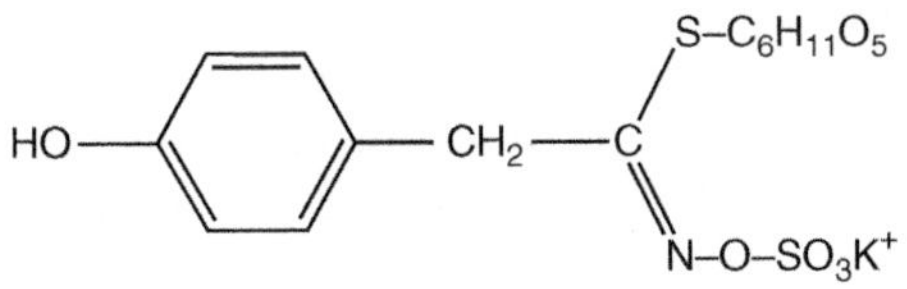

sinalbin from white mustard seed

Therapeutic actions

The main use for the brassicas in general is culinary. In commerce black mustard has largely been replaced by brown mustard (*Brassica juncea*) in recent times, since the latter species is better adapted to mechanical harvesting (Tucker and DeBaggio, 2000). Apart from providing a rich source of nutrients, black and brown mustard seeds act to stimulate appetite and digestion. Mustard seeds are rich in mucilage and essential fatty acids as well as glucosinolates. Seed oils act as rubefacients or irritants when applied topically, causing local vasodilation. Mustard poultices have been used historically to break up congestion in the lungs and bronchioles, though care must be taken not to induce inflammatory skin reactions.

Table 4.3. Glucosinolates and their precursors

Glycoside	Aglycone	Precursor	Herb
Sinigrin	Allyl isothiocyanate	Methionine	*Brassica nigra, Brassica juncea, Brassica oleracea*
Sinalbin	*p*-Hydroxybenzyl isothiocyanate	Tyrosine	*Sinapis alba*
Gluconasturtiin	Phenylethyl isothiocyanate	Phenylalanine	*Armoracia rusticana, Rorippa nasturtium-aquaticum*
Glucotropaeolin	Benzyl isothiocyanate	Phenylalanine	*Tropaeolum majus, Lepidium sativum*
Glucobrassicin	3-Indolylmethyl isothiocyanate	Tryptophan	*B. oleracea, Brassica rapa, Brassica napus*
Progoitrin	2-Hydroxy-3-butenyl isothiocyanate	Tryptophan	*B. oleracea*
Glucoraphanin	Sulforaphane	Methionine	*B. oleracea, B. rapa*

Taken internally the glucosinolate herbs are effective decongestants for sinus and bronchial conditions (e.g. horseradish tablets, often combined with garlic), while also acting to stimulate digestion and circulation. Nasturtium is traditionally used for acute bronchitis and in dermatology for skin rashes, mild burns and dandruff (Bruneton, 1995). Like many sulfur-containing compounds, glucosinolates exhibit antibiotic effects.

Dietary glucosinolates found in all brassicas (including kale, turnips and radishes) are inactive until being hydrolysed following consumption to aglycones such as isothiocyanates (sinigrin) and indoles (**glucobrassicin**). They are preserved after cooking. These metabolites are regarded as being among the most significant cancer preventative compounds in the human diet. Several mechanisms for cancer prevention have been demonstrated, including induction of Phase 1 and 2 detoxification enzymes, apoptosis (programmed cell death) of cancer cells and suppression of oestrogen-dependent cancer cells (Cartier and Valesco, 2008).

Glucosinolates from mustard seeds are potent pesticides for control of weeds, nematodes, insects and fungi. Currently they are being tested with the objective of being proven suitable for use in organically certified farms (Popova *et al.*, 2017).

Toxicology

Volatile oil of mustard seed consists purely of glucosinolates, and it should never be used internally or externally. The oil is classified as severely toxic and irritant to skin and mucous membranes, and the oral LD_{50} (lethal dose required to kill 50% of a tested population after a specified test duration) is very low at 0.15 (Tisserand and Balacs, 1995). Inhalation of the oil induces severe irritation of the eyes and nasal membranes. Essential oil derived from horseradish root (*Armoracia rusticana*, Brassicaceae) is equally toxic.

Powdered black mustard seed is the form in which fomentations and other topical medications are traditionally prepared. Severe burning can occur after 15–30 minutes application of the pure powder, but the potency can be reduced by mixing with a carrier such as corn starch. Mustard therapy is contraindicated in cases of varicose veins and other circulatory disorders (Bissett, 1994).

Ingestion of large quantities of glucosinolates can produce irritant poisoning; symptoms include burning, abdominal pain, vomiting and in severe cases loss of consciousness and death (Wink and van Wyk, 2008).

Hypothyroid individuals should avoid overconsumption of glucosinolate herbs and foods. Progoitrins in brassicas are hydrolysed to oxazolidine-2-thione, a goitrogenic compound that blocks iodine synthesis. This is a major problem for livestock being fed meal or 'cake' from mustard and rapeseed residue left over from oil seed production. There are few reports of goitrogenic effects in humans (Cartier and Valesco, 2008).

References

Adzet, T. and Camarasa, J. (1988) Pharmacokinetics of polyphenolic compounds. In: Cracker, L.E. and Simon, J.E. (eds) *Herbs, Spices and Medicinal Plants*, Vol. 3. Oryx Press, Phoenix, Arizona, pp. 25–47.

Bissett, N.G. (ed.) (1994) *Herbal Drugs and Phytopharmaceuticals*. Medpharm Publications, Stuttgart, Germany.

Bolarinwa, I.F., Oke, M.O., Olaniyan, S.T. and Ajala, A.S. (2016) A review of cyanogenic glycosides in edible plants. In: Larramendy, M. and Soloneski, S. (eds) *Toxicology: New Aspects to This Scientific Conundrum*. IntechOpen Limited, London, pp. 179–192. doi: 10.5772/64886

Bruneton, J. (1995) *Pharmacognosy, Phytochemistry, Medicinal Plants*. Lavoisier Publications, Paris.

Cartier, M.E. and Valesco, P. (2008) Glucosinolates in *Brassica* foods: bioavailability in food and significance for human health. *Phytochemical Review* 7, 213–229. doi: 10.1007/s11101-007-9072-2

Clar, E. (1964) *Polycyclic Hydrocarbons*, Vol. 1. Springer, Heidelberg, Germany.

Cometa, F., Nicoletti, T.M. and Pieretti, S. (1993) Phenylpropanoid glycosides, distribution and pharmacological activity. *Fitoterapia* 64, 195–217.

Demirezer, L.O. and Uzun, M. (2016) Determination of sun protection factor (SPF) of *Rumex crispus* and main anthraquinones. *Planta Medica* 82(S01), S1–S381. doi: 10.1055/s-0036-1596459

Dewick, P.M. (2009) *Medicinal Natural Products: a Biosynthetic Approach*, 3rd edn. Wiley, Chichester, UK. doi: 10.1002/9780470742761

Facino, R., Carina, M., Aldini, G., Saibene, L., Pietta, P. and Mauri, P. (1995) Echinacoside and caffeoly conjugates protect collagen from free radical-induced degradation. *Planta Medica* 61, 510–514. doi: 10.1055/s-2006-959359

Gardner, Z. and McGuffin, M. (2013) *Botanical Safety Handbook*, 2nd edn. American Herbal Products Association (AHPA). CRC Press, Boca Raton, Florida.

Günaydin, K., Topçu, G. and Ion, R.M. (2002) 1,5-Dihydroxyanthraquinones and an anthrone from roots of *Rumex crispus*. *Natural Product Letters* 16(1), 65–70. doi: 10.1080/1057563029001/4872

Herbert, J.M., Maffrand, J.P., Taoubi, K., Augereau, J.M., Fouraste, J. and Gleye, J. (1991) Verbascoside isolated from *Lantana camara*, an inhibitor of protein kinase C. *Journal of National Products* 54, 1595–1600. doi: 10.1021/np50078a016

Klimek, B., Olszewska, M.A. and Tokar, M. (2010) Simultaneous determination of flavonoids and phenylethanoids in the flowers of *Verbascum densiflorum* and *V. phlomoides* by high-performance liquid chromatography. *Phytochemical Analysis* 21, 150–156. doi: 10.1002/pca.1171

Macleod, A.J. (1976) Volatile flavour compounds of the Cruciferae. In: Vaughan, J.G., Macleod, A.J. and Jones, B.M.G. (eds) *The Biology and Chemistry of the Cruciferae*. Academic Press, London, pp. 307–330.

Milazzo, S., Lejeune, S. and Ernst, E. (2007) Laetrile for cancer: a systematic review of the clinical evidence. *Support Care Cancer* 15(6), 583–595. doi: 10.1007/s00520-006-0168-9

Pan, J., Yuan, C.S., Lin, C.J., Jia, Z.J. and Zheng, R.L. (2003) Pharmacological activities and mechanisms of natural phenylpropanoid glycosides. *Pharmazie* 58(11), 767–775. doi: 10.1002/chin.200405273

Panossian, A., Wikman, G. and Wagner, H. (1999) Plant adaptogens III. Earlier and more recent aspects and concepts on their mode of action. *Phytomedicine* 6, 287–300. doi: 10.1016/S0944-7113(99)80023-3

Panossian, A., Wikman, G. and Sarris, J. (2010) Rosenroot (*Rhodiola rosea*): traditional use, chemical composition, pharmacology and clinical efficacy. *Phytomedicine* 17, 481–493. doi: 10.1016/j.phymed.2010.02.002

Pieretti, S., Di Giannuario, A., Capasso, A. and Nicoletti, M. (1992) Pharmacological effects of phenylpropanoid glycosides from *Orobanche hederae*. *Phytotherapy Research* 6(2), 83–93. https://doi.org/10.1002/ptr.2650060208

Popova, I.E., Dubie, J.S. and Morra, M.J. (2017) Optimization of hydrolysis conditions for release of biopesticides from glucosinolates in *Brassica juncea* and *Sinapis alba* seed meal extracts. *Industrial Crops and Products* 97, 354–359. http://dx.doi.org/10.1016/j.indcrop.2016.12.041

Sasaki, H., Nishimura, H., Morota, T., Chin, M., Mitsuhashi, H., *et al.* (1989) Immunosuppressive principles of *Rehmannia glutinous* var. *hueichingensis*. *Planta Medica* 55, 458–462. doi: 10.1055/s-2006-962064

Tisserand, R. and Balacs, T. (1995) *Essential Oil Safety*. Churchill Livingstone, Edinburgh, UK.

Tucker, A.O. and DeBaggio, T. (2000) *The Big Book of Herbs*. Interweave Press, Fort Collins, Colorado.

Wagner, H., Norr, H. and Winteroff, H. (1994) Plant adaptogens. *Phytomedicine* 1, 63–76. doi: 10.1016/S0944-7113(11)80025-5

Wink, M. and van Wyk, B. (2008) *Mind-altering and Poisonous Plants of the World*. Briza Publications, Pretoria, South Africa.

Zhuang, W., Yue, L., Dang, X., Chen, F., Gong, Y., *et al.* (2019) Rosenroot (*Rhodiola*): potential applications in aging-related diseases. *Aging and Disease* 10(1), 134–146. https://doi.org/10.14336/AD.2018.0511

Terpenes

5

Introduction

Terpenes and terpenoids (terpenes with oxygen) comprise one of the most important groups of active compounds in plants with over 20,000 known structures. Terpenoids are isoprenoids, since their structures are built upon isoprene (five-carbon) units, containing two unsaturated bonds. Most are synthesized from the five-carbon precursors, isopentenyl pyrophosphate (IPP) and its isomer dimethylallyl diphosphate (DMAPP), via one of two biosynthetic pathways. In the majority of plants, fungi and mammals they are synthesized from acetate, via the mevalonic acid pathway, while in bacteria and some plants (one being *Ginkgo biloba*, Ginkgoaceae) they come from pyruvate via the methylerythritol 4-phosphate (MEP) pathway (Kuzuyama and Seto, 2012).

$$H_2C = C(CH_3) - CH = CH_2$$

isoprene – C_5H_8

During the formation of the terpenes, the isoprene units are most commonly linked in a head-to-tail fashion. The number of units incorporated into a particular terpene serves as a basis for the classification of these compounds, as shown in Table 5.1.

The process of terpene biosynthesis involves vast numbers of enzymes, mainly of the terpene synthase gene family or TPS. Upon formation of the immediate precursors to mono- and sesquiterpenes, reactions with TPS lead to the formation of C_{10} monoterpenes and C_{15} sesquiterpenes, while formation of some terpenoids also requires cytochrome P450 enzymes (Cheng *et al.*, 2007; Chen *et al.*, 2011; Padovan *et al.*, 2014).

DOI: 10.1079/9781789243079.0005

Table 5.1. Classification of terpenes

Category	Molecular formula	Components of:
Isoprene	C_5H_8	
Terpene	$(C_5H_8)n$	
Monoterpene	$C_{10}H_{16}$	Essential oils (e.g. menthol, iridoids)
Sesquiterpene	$C_{15}H_{24}$	Bitter principles, especially sesquiterpene lactones
Diterpene	$C_{20}H_{32}$	Resin acids, bitter principles
Triterpene	$C_{30}H_{48}$	Saponins, steroids
Tetraterpene	$C_{40}H_{64}$	Carotenoids
Polyterpene	$(C_5H_8)n$	Rubber

It should be noted that isoprene units – generally a dimethylallyl substituent – are often present in unrelated molecules such as flavonoids and alkaloids. The term 'prenyl' indicates an isoprenyl substitute, and molecules that are 'prenylated' in this fashion are rendered more lipophilic. Examples of prenylated molecules will occur in later chapters.

Monoterpenes

These are the major class of chemical compounds found in essential oils. Among the most widely occurring monoterpene oils are 1,8-cineole from *Eucalyptus* spp. and pinene from *Pinus* spp. These are covered in the essential oil chapter (Chapter 8, this volume).

Monoterpene biosynthesis involves condensation of two C_5 precursors giving rise to geranyl pyrophosphate (GPP), a C_{10} intermediate from which monoterpenes are derived. NerylPP and linalylPP, isomers of GPP, are reactive molecules capable of cyclization to form the methane ring system. Depending on the specific monoterpene synthases present in the plant, by giving up their phosphate groups these molecules act as substrates for a host of linear, cyclic and bicyclic molecular structures – the monoterpenoids (Dewick, 2009).

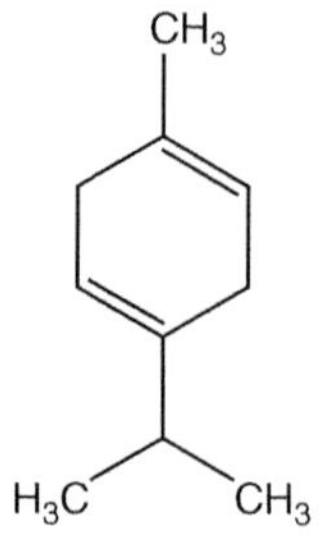

γ-terpinene – a monoterpene

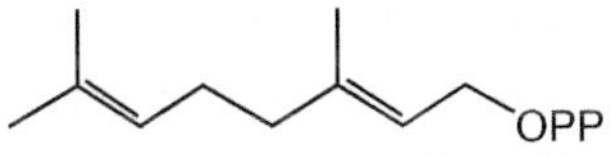

geranyl pyrophosphate (GPP)

Condensation of GPP with C_5 precursors gives rise to diterpenes, and the other classes of terpenoids are similarly formed.

Paeoniflorin, a monoterpene glucoside with a cage-like structure bonded to a benzene ring, is a major constituent found in the Chinese herb, *Paeonia lactiflora* (Paeoniaceae). The herb is anti-inflammatory, sedative, antipyretic and antispasmodic (Huang, 1993). Research findings also support antiallergy activity based on IgE (immunoglobulin E)-induced anaphylaxis and scratching behaviour *in vivo* (Lee et al., 2008).

Pyrethrins – irregular monoterpenoids

A group of terpenoids found in species of *Chrysanthemum* and *Tanacetum* (in the Asteraceae family) contain compounds such as **chrysanthemic acid** in which the coupling of isoprene units is not in the usual head-to-tail style, hence the 'irregular' designation. When esterfied (bonded to a complex alcohol) **pyrethrins**, source of the widely used insecticidal products, are formed.

chrysanthemic acid

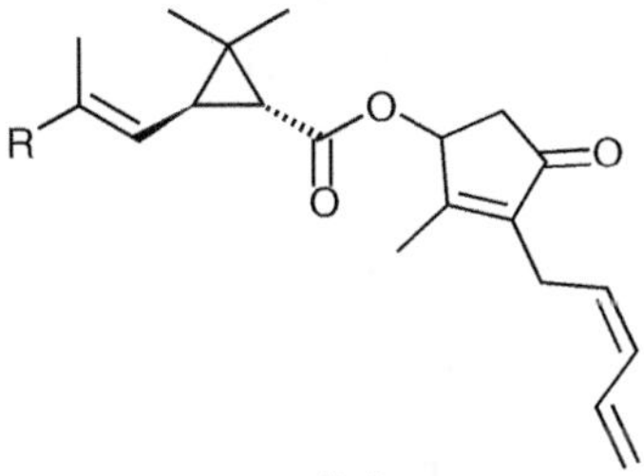

pyrethrin

Iridoids

Iridoids are synthesized through the mevalonic acid pathway and are technically known as cyclopentan-[*c*]-pyran monoterpenoids. The precursor monoterpene – geraniol – undergoes a series of oxidation, hydroxylation and glycosylation reactions, leading to the formation of **loganin**, an iridoid glycoside.

The name iridoid is derived from the common Australian meat ant *Iridomyrmex detectus*, from which it was first detected in 1956 (Sticher, 1977). Iridoid glycosides are derived from plants belonging to many families, most notably the Rubiaceae, Lamiaceae, Scrophulariaceae and Gentianaceae. The first iridoid glycoside to be identified was **asperuloside** from woodruff (*Asperula odorata*, Rubiaceae). Other compounds of therapeutic significance include **aucubin** from plantain (*Plantago* spp., Plantaginaceae) and eyebright (*Euphrasia officinalis*, Orobanchaceae), **catapol** also from *E. officinalis* and loganin from bogbean (*Menyanthes* spp., Menyanthaceae).

Non-glycosidic iridoids include the sedative valepotriates found in valerian rootlets (*Valeriana* spp., Valerianaceae). **Valtrate**, an epoxy triester is the principle component – up to 80% – but there are many others. Valepotriates are unstable molecules, and readily decompose while the rootlets are drying to give off isovaleric acid, the characteristic 'dirty socks' odour associated with valerian (Dewick, 2009). Valepotriates also decompose in water or ethanol, the main metabolite being the baldrinals, aldehydes which are still biologically active (Bos *et al.*, 2002).

Another non-glycosidic iridoid is **nepetalactone**, the volatile component of essential oil of catnip (*Nepeta cataria*, Lamiaceae), responsible for attracting cats to the plant.

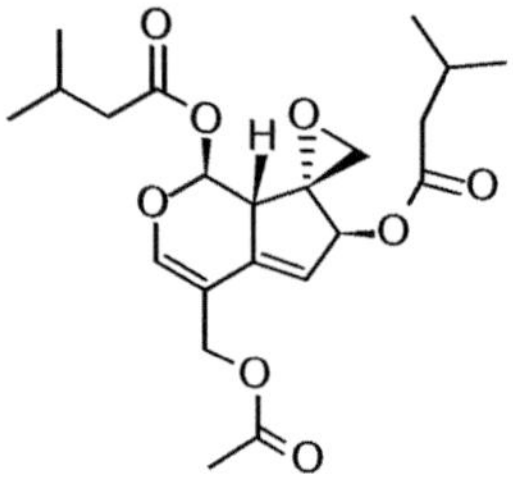

valtrate – an epoxyiridoid ester

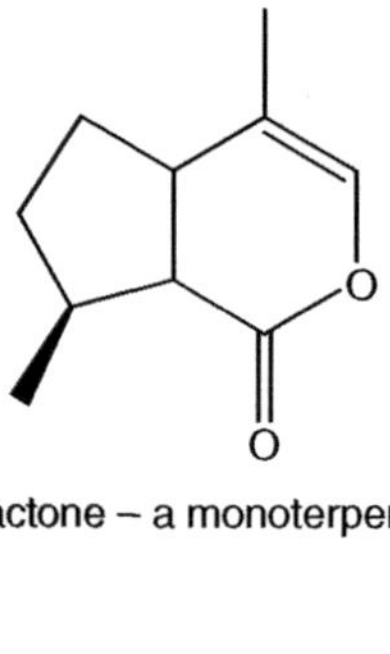

nepetalactone – a monoterpene iridoid

Secoiridoids
These glycosides are formed by the opening of the five-carbon ring of the iridoid loganin. They include **amarogentin** and **gentiopicroside** from gentian (*Gentiana* spp., Gentianaceae), **picroliv** from *Picrorhiza kurroa* (Plantaginaceae) and

oleuropein from olive leaves (*Olea europaea*, Oleaceae). The Gentianaceae family is an extremely rich source of these glycosides: for a comprehensive review of these, refer to Rodriguez *et al.* (1998).

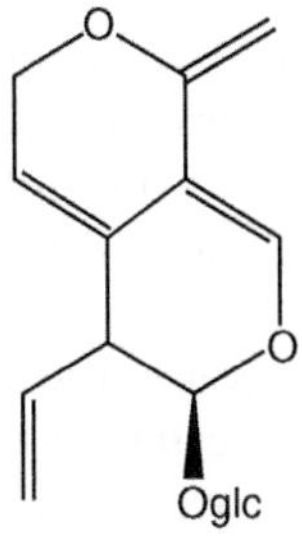

gentiopicroside – a secoiridoid glycoside

Harpagide, an anti-inflammatory compound found in devil's claw (*Harpagophytum procumbens*, Pedaliaceae), and figwort (*Scrophularia nodosa*, Scrophulariaceae), is a decarboxylated iridoid glycoside. **Harpagoside**, the cinnamoyl ester of harpagide, contains a phenylpropanoid moiety. *H. procumbens*, a tuberous plant found in the Kalahari Desert region of Southern Africa, is a traditional medicine for fevers and malaria as well as for rheumatic disorders. It contains at least 12 iridoid glycosides along with several triterpenoids (Qi *et al.*, 2010).

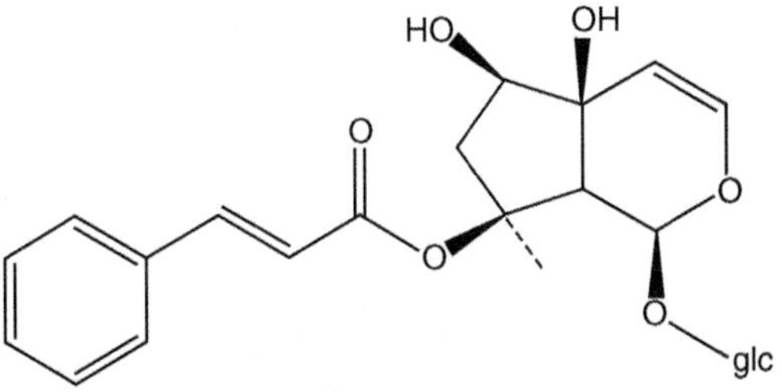

harpagoside

Therapeutic actions of iridoids

Iridoids are the most bitter of all plant compounds, often responsible for the so-called 'bitter principle'. On a scale for bitter value devised by Wagner and Vaserian (described in Sticher, 1977), amarogentin and related secoiridoids were the most bitter of all compounds tested. The taste is perceptible at a dilution of 1 part in 50,000.

Various iridoids have been identified as antimicrobial, laxative, choleretic and hepatoprotective – especially picroliv (Sticher, 1977; Visen *et al.*, 1993). While anti-inflammatory activity has been identified in aucubin and others, in most cases it is relatively weak. In the case of harpagoside this may depend on the degree of hydrolysis that occurs in the gut since the aglycone (harpagogenin) was found to be less active than the glycoside itself (Recio *et al.*, 1994).

One of the richest sources of iridoid glycosides is the Chinese herb, *Rehmannia glutinosa* (Orabanchaceae), which is also highly regarded in Western herbalism. Along with the common iridoids acubin and catapol, in this plant there are more complex molecules in which the iridoids are bonded to cinnamic, vanillic and syringic acids, a lignan, norcarotinoid esters, sesquiterpenoids and a rare example of an iridoid glycoside dimer (Liu *et al.*, 2012). The species is commonly used as an anti-inflammatory and antianaemic agent, while several of these iridoids demonstrated hepatoprotective action *in vitro* (Liu *et al.*, 2012).

The olive fruit and leaf (*O. europaea*, Oleaceae) contain the secoiridoids oleuropein and ligstroside, these are esters of the phenol hydroxytyrosol, the potent antioxidant also found in white wine. Oleuropein may occur in numerous other plants of the Oleaceae, including ash (*Fraxinus* spp.), privet (*Ligustrum* spp.), lilac (*Syringa* spp.) and fringe tree (*Chionanthus virginicus*) (Omar, 2010).

Pharmokinetic investigations indicate that oleuropein is highly bioavailable, with the main metabolites being hydroxytyrosol and tyrosol (Vissers *et al.*, 2002). These metabolites are potent antioxidants, with cardioprotective and anticancer effects as noted in Chapter 2, this volume. Oleuropein and related compounds are responsible for much of the antimicrobial action of olive leaf. Additional pharmacological actions include antioxidant, anti-inflammatory, anti-atherogenic, anticancer and antiviral (Omar, 2010).

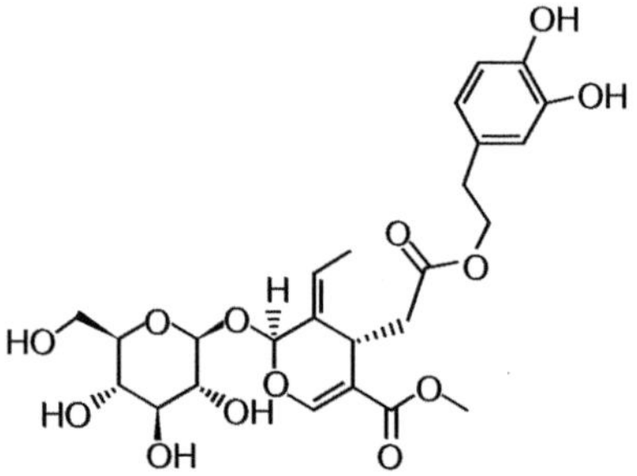

oleuropein

Many molecular targets have been identified by which olive phytochemicals, particularly oleuropein, may act as chemopreventive agents in cancer. These include HER2, epigenetic modifications, interference with the MAPK (mitogen-activated protein kinase) pathway, modulation of apoptosis and the phosphatidylinositol 3-kinase (PI3K)/Akt signalling axis (Farooqi *et al.*, 2017). Most attention has been paid to breast cancer, where oleuropein inhibits molecules responsible for metastatic effects (i.e. the spread of cancer cells from the original tumour) (Hassan *et al.*, 2012). Also promising are findings that indicate oleuropein and lignans from olive downregulate HER2 protein expression, *HER2* being the gene most strongly linked to the progression of breast cancer (Menendez *et al.*, 2008).

Sesquiterpenes

These C_{15} compounds occur mainly as ingredients of essential oils or as γ-lactones.

They are thought to have evolved as phytoalexins or antifeedants, compounds synthesized by plants as a response to fungal attack and herbivore grazing (Cronquist, 1988; Bruneton, 1995). They are synthesized via mevalonic acid, the immediate precursor being farnesyl pyrophosphate (FPP). In essential oils they are usually found in combination with monoterpenes, although they have higher melting points.

Flowers of German chamomile (*Matricaria recutita*, Asteraceae) are rich in sesquiterpenes, most notable the alcohol **bisabolol** and **bisabolol oxides A** and **B**, compounds with potent anti-inflammatory and wound-healing actions. Another sesquiterpene, **matricin**, is also anti-inflammatory, however, distillation of the flowers converts the compound to the blue-coloured **chamazulene**, a less active compound. Carbon dioxide (CO_2) extraction of the flowers is preferred by some aromatherapists, as the more potent anti-inflammatory compound matricin is not converted to chamazulene by this process (Guba, 2018). The essential oil of Roman chamomile (*Chamaemelum nobile*, Asteraceae) also contains chamazulene, however, only at low levels, and the herb shouldn't be regarded as being equivalent to *M. recutita*.

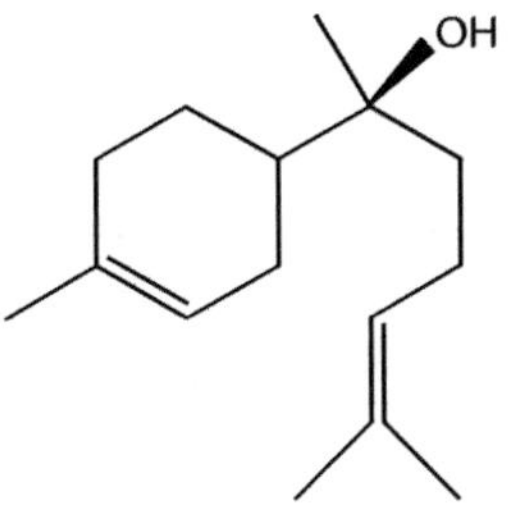

bisabolol – a sesquiterpene

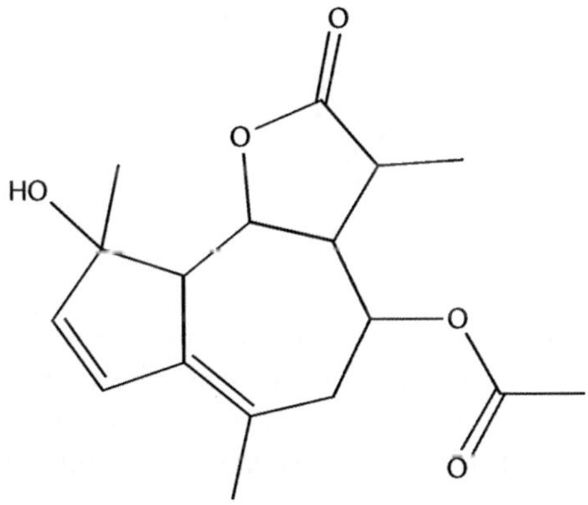

matricin

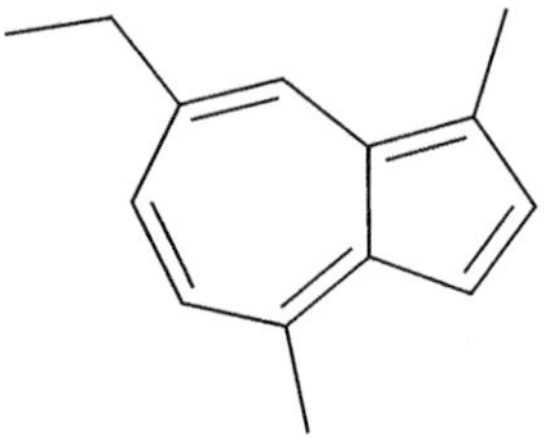

chamazulene

Gossypol from the Levant cotton plant (*Gossypium herbaceum*, Malvaceae) is a sesquiterpene dimer, in which the isoprene units are arranged into two bonded aromatic structures. Gossypol has been used in China as an antifertility agent in men. It is thought to act primarily on the testicular mitochondria, by inhibiting calcium ion (Ca^{2+}) uptake at the presynaptic endings (Huang, 1993). Despite the high efficacy (> 99%) of gossypol, its long-term use as an infertility agent is limited by undesirable side effects which include irreversible infertility (Hamburger and Hostettmann, 1991). However, it appears that most of the toxicity is associated with the (+)-isomer of gossypol found in some other species of *Gossypium*, whereas the contraceptive action is caused by the (–)-isomer found only in *G. herbaceum* (Dewick, 2009).

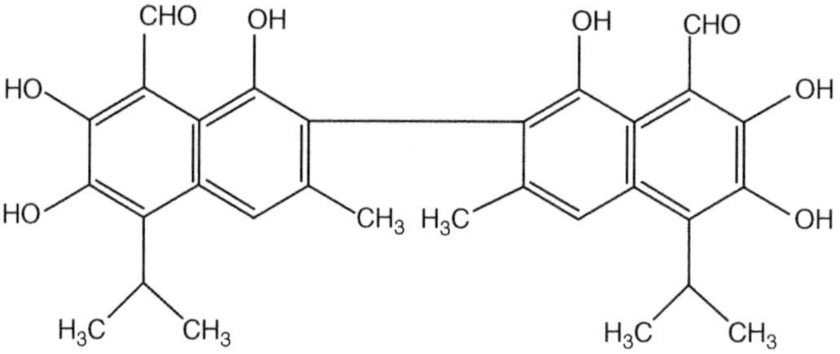

gossypol – a sesquiterpene dimer

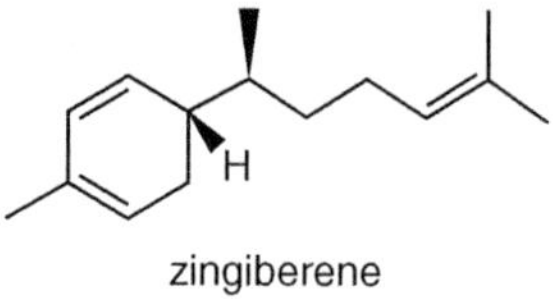

zingiberene

Sesquiterpenes are significant constituents of myrrh resin (*Commiphora molmol*, Burseraceae), responsible for the local anaesthetic, antibacterial and antifungal properties of the resin (Dolara *et al.*, 2000). Ginger (*Zingiber officinale*, Zingiberaceae) is another resinous plant containing the sesquiterpenes

zingerbene, β-**bisabolene** and β-**phellandrene**, which are major flavour compounds, also contributing to the taste, odour and therapeutic actions of ginger (van Wyk, 2013).

Sesquiterpene lactones

Over 5000 different sesquiterpene lactones (SLs) are known, the vast majority occurring in the Asteraceae family. They often occur as mixtures of several related compounds, and tend to concentrate in leaves and flowers. Structurally they consist of one and a half terpenes (or three isoprene units) attached to a lactone ring (γ-lactone). The common precursor of SLs is thought to be the sesquiterpene **germacrene**, synthesized in one step from FPP, then further oxidized to form **costunolide**, the direct precursor of the SL parthenolide (Majdi *et al.*, 2011).

Many of their names end in the suffix 'olide' indicating the presence of the lactone group. Being such a large group, the lactones are classified into several main subgroups according to their ring structures (Rodriguez *et al.*, 1976; Harborne and Baxter, 1993; Dewick, 2009):

- eudesmanolides – two fused six-membered rings;
- guaianolides – a five-membered ring fused to a seven, methyl substituent at C-4;
- pseudoguaianolides – as above but with methyl substituent at C-5;
- germacranolides – a ten-membered ring. This is the precursor to parthenolide; and
- humulyl cation – an eleven-membered ring. This is the precursor to β-caryophyllene.

Some Asteraceous medicinal plants and their main sesquiterpene ingredient are listed in Table 5.2.

Table 5.2. Sesquiterpene lactones (SLs) from the Asteraceae

Species	SL
Achillea millefolium	Achillin
Inula helenium	Alantolactone
Artemisia annua	Artemisinin
Artemisia absinthium	Absinthin
Cnicus benedictus	Cnicin
Arnica montana	Helenalin
Tanacetum parthenium	Parthenolide
Tanacetum vulgare	Tanacetin
Cichorium intybus	Lactucin

Apart from their benefit as digestive bitters, many of these compounds have anti-inflammatory, antimicrobial and antiprotozoal actions.

Parthenolide from feverfew (*Tanacetum parthenium*, Asteraceae) has been widely acclaimed for its benefits in treatment and prevention of migraine headaches.

The germacranolide-type SL, distinguished by an electrophilic epoxide group and a methylene group on the lactone ring, is mainly concentrated in glandular trichomes (hairs) of the disc florets of feverfew, with lesser amounts in the ray florets and leaves (Majdi *et al.*, 2011). These structural features are alkylating sites which enable interactions with thiol groups in proteins and DNA, producing cytotoxic effects that may be beneficial in cancer treatment, while also capable of inducing allergic reactions by formation of antigens (Dewick, 2009; Chadwick *et al.*, 2013). Anti-inflammatory and migraine-relieving effects are linked to inhibition of platelet aggregation, as well as the release of histamine and production of prostaglandins. Significant antipyretic, analgesic and uterine stimulating actions have been demonstrated for feverfew *in vivo*, though these effects may not all be due to SLs (Rateb *et al.*, 2007). Other active constituents of feverfew include flavonoids and the essential oil, over half of which is made up of the monoterpene ketone camphor.

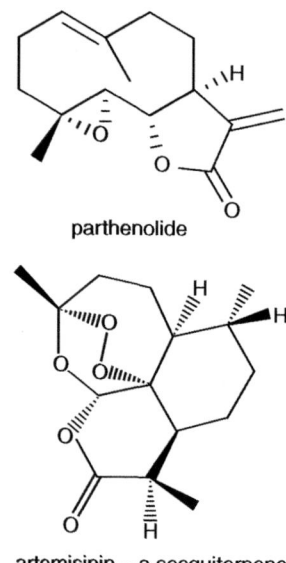

parthenolide

artemisinin – a sesquiterpene
lactone endoperoxide

Artemisinin, from sweet Annie (*Artemisia annua*, Asteraceae), is widely used to treat malaria. The presence of the endoperoxide group attached to the guaianolide-type sesquiterpene structure is responsible for the antiparasitic action. *A. annua* and *Artemisia absinthium* (wormwood) and their SLs are also effective against various forms of parasitism at low dose, including food-borne trematodiasis and schistosomiasis – diseases that afflict millions of people

(Ferreira *et al.*, 2011). While artemisinin has become a standard treatment for malaria in many parts of the world, issues of drug resistance have emerged, particularly in Asia (Dondorp *et al.*, 2009; Fairhurst *et al.*, 2012). This problem could be addressed by using more traditional preparations such as extracts or teas made from the whole *A. annua* herb, instead of the isolated artemisinin (Spelman, 2009). At the other end of the scale, biosynthesis of artemisinin has been achieved starting with the monoterpene citronellal, while production of the molecule using genetically modified species has also been investigated (Fraga, 2012).

The roots of elecampane (*Inula helenium*, Asteraceae) have long been used to treat lower respiratory infections. The roots are rich in SLs, mainly of the eudesmanolide type. **Alantolactone** and several derivatives including isoalantolactone and epoxyalantalactone showed antibacterial activity against several human pathogens (Boatto *et al.*, 1994; Jiang *et al.*, 2011). Alantolactone and related SLs acting as electrophiles can interact with genes that metabolize Phase 2 detoxifying enzymes, providing validation for the use of elecampane root for chemopreventation against cancer (Seo *et al.*, 2009).

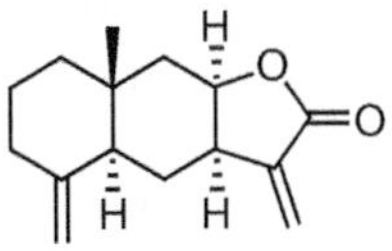

isoalantolactone

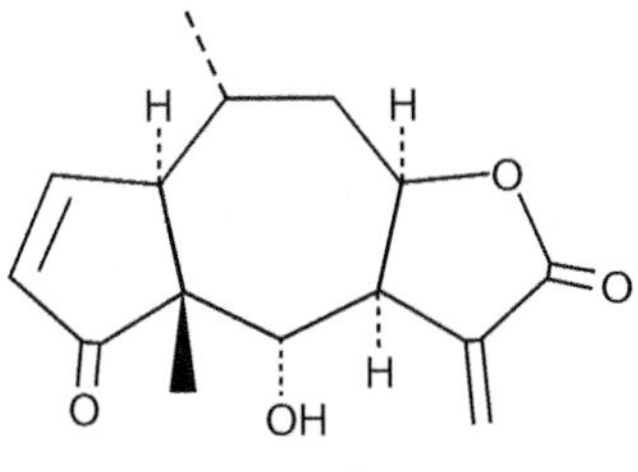

helenalin

Arnica montana (Asteraceae) is a widely used anti-inflammatory herb, whose main active constituents are the SL helenalin and related compounds. One of these, **dihydrohelenalin** – and particularly its ester derivatives – reduce inflammation via inhibition of cytokines TNF-α (tumour necrosis factor-alpha) and IL-1 (interleukin-1), by regulation of their transcription factors, in a similar way to glucocorticoids (Klaas *et al.*, 2001).

Allergic contact dermatitis
Numerous SL-containing plants of the Asteraceae family are known to cause contact dermatitis in humans. In one case a patient developed acute dermatitis of

the right hand with severe blistering following a single application of Arnica tincture (Hormann and Korting, 1995). The presence of an α-methylene group in the chemical structure is thought to be a prerequisite for production of dermatitis (Rodriguez *et al.*, 1976). Essential oil of costus (*Saussurea costus*, Asteraceae), in which costus lactones are the major components, was formerly used in perfumery; however, it has been identified as a powerful sensitizer and is now rarely used (Tisserand and Balacs, 1995).

A toxic SL, anisatin, a convulsive, is found in *Illicium anisatum* (Schisandraceae) a close relative and potential adulterant of the popular culinary star anise (*Illicium verum*) (van Wyk and Wink, 2014).

Diterpenes

These are among the most bitter tasting of all terpenoid compounds, responsible for the acclaimed stomachic and tonic properties of herbs such as **marrubiin** in horehound (*Marrubium vulgare*, Lamiaceae), and **calumbin** from calumba (*Jateorhiza palmata*, Menispermaceae). They occur as acyclic, cyclic, bi-, tri- and tetra-cyclic compounds (referring to the number of carbon rings present), while many are also lactones. Some are chiral molecules, indicated by the prefix 'ent' (for enantiomers).

Diterpenes tend to be abundant in the Lamiaceae family, for example *Salvia officinalis*, or common sage, which has antiviral diterpenes (Tada *et al.*, 1994) and *Salvia divinorum* contains the hallucinogen **salvinorin A. Forskolin**, a diterpene oxide, is the major principle of *Plectranthus forskohlii* (syn. *Coleus forskohlii*, Lamiaceae). Forskolin has demonstrated a range of activities, including vasodilatory, antihypertensive, bronchodilatory, positive inotropic action on the heart, decreases intraocular pressure, and inhibition of platelet aggregation (Hamburger and Hostettmann, 1991; Bruneton, 1995).

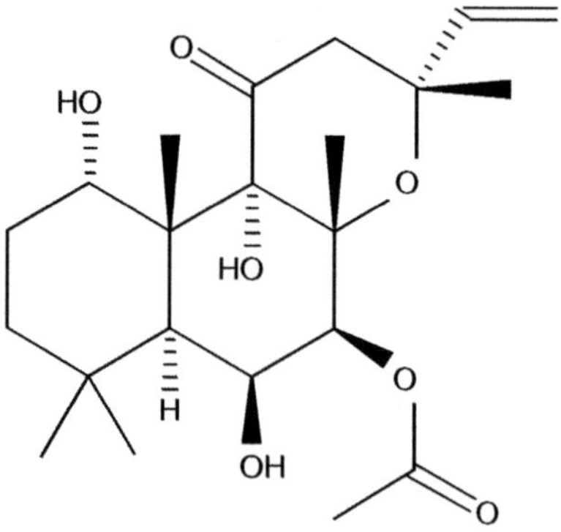

forskolin – a labdane diterpene oxide

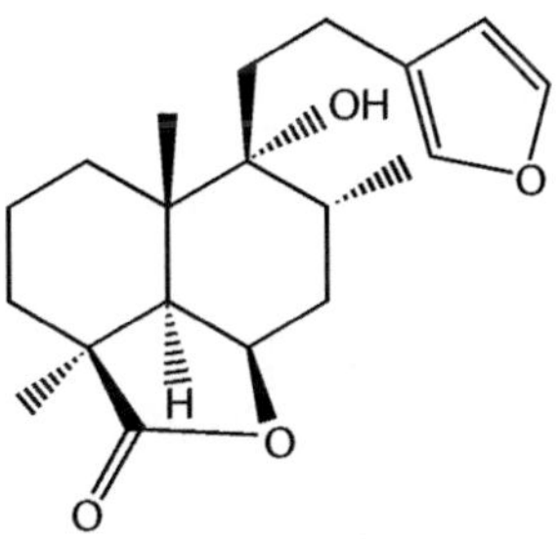

marrubiin – a furanic
labdane diterpene

Paclitaxel (taxol) is a complex diterpene ester from the yew tree, especially the Pacific yew (*Taxus brevifolia*, Taxaceae). Taxol and its derivatives are used as prescription drugs for cervical and breast tumours (Lenaz and De Furia, 1993). It has an antimitotic effect, associated with an interaction of taxol with the tubulin-microtubule system, thereby blocking proliferation of cancer cells (Hamburger and Hostettmann, 1991; Dewick, 2009). **Taxol®** is also the registered name of a cancer-treating drug.

The Indian herb *Andrographis paniculata* (Acanthaceae) contains diterpene lactones, glucosides and dimers including **andrographolide** (Das *et al.*, 2010). Hepatoprotective activity has been demonstrated for andrographolide, supporting the traditional use for liver and digestive disorders (Chander *et al.*, 1995). Andrographolide and its derivatives provide substantial immune-enhancing effects, verified by clinical studies in bacterial and respiratory infections (Melchior *et al.*, 2000). They are also being investigated as anticancer agents (Das *et al.*, 2010).

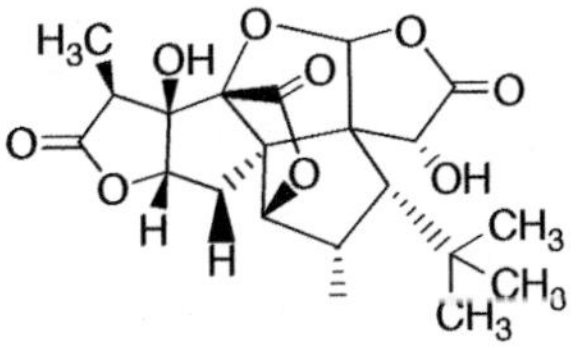

ginkgolide A – a diterpene trilactone

Ginkgo biloba (Ginkgoaceae) diterpenes or ginkgolides are trilactones (i.e. they contain three lactone rings). **Ginkgolides** are PAF inhibitors (i.e. inhibitors of platelet activation factor (PAF), a signalling molecule that not only thickens blood, but which acts as an inflammation mediator). **Bilobide** has

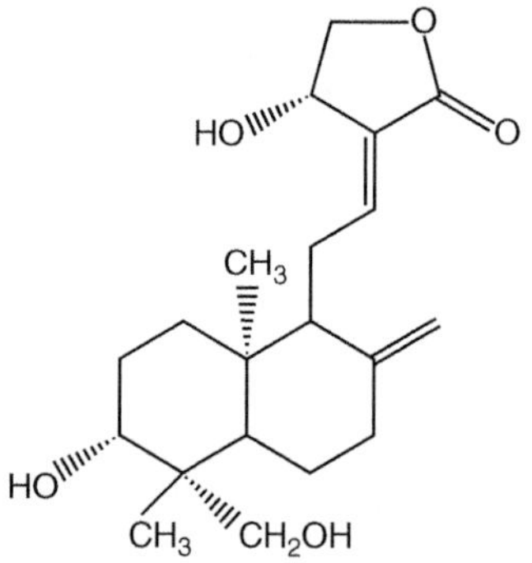

androgragpholide – a diterpene lactone

the same structure but with one less carbon ring, it is classified with the sesquiterpenes. Bilobide is neuroprotective. Both ginkgolides and bilobide mediate the neurotransmitters GABA (γ-aminobutyric acid), glycol and 5HT (5-hydroxytryptamine or serotonin), bilobide being the most potent (Huang *et al.*, 2004).

The Australian hop bush (*Dodonaea viscosa*, Sapindaceae) which is also distributed across parts of Oceania, Africa, Southern Asia and Central America, is a rich source of pharmacologically active diterpenoids. Ent-labdanes such as **hautriwaic acid** have been shown to produce spasmolytic effects on isolated guinea-pig ileum, that involve interference with calcium influx into the smooth muscle cells (Mata *et al.*, 1999). Hautriwaic acid also demonstrated anti-inflammatory activity *in vivo* (Salinas-Sánchez *et al.*, 2012).

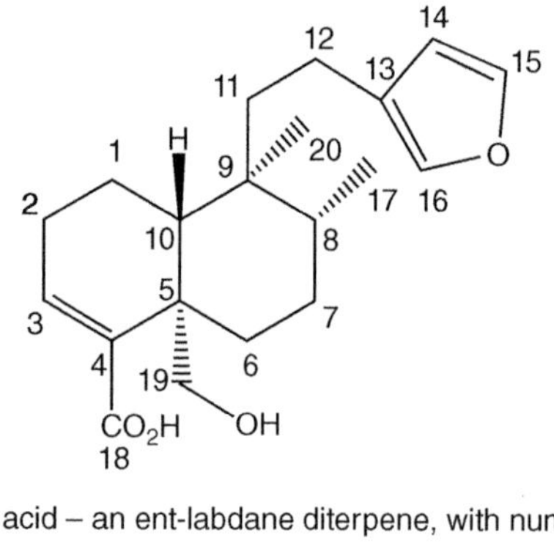

hautriwaic acid – an ent-labdane diterpene, with numbered carbons

Stevioside and **rebaudioside** are sweet-tasting diterpenes from *Stevia rebaudiana* (Asteraceae). The glycosides have demonstrated a range of biological activities, including hypotensive, antiviral and gastroprotective *in vivo*, while also being investigated as a treatment for type 2 diabetes, a traditional use for this species (Ferriera *et al.*, 2006; Madan *et al.*, 2010).

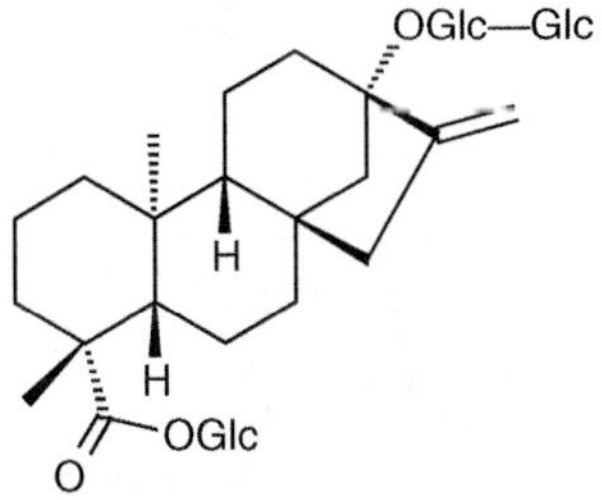

stevioside – a diterpene glycoside

Table 5.3 lists major medicinal plants that contain diterpenes.

Toxicity of diterpenes

While there is generally little toxicity associated with diterpenes, there are a group of diterpene phorbal esters (e.g. **myristoylphorbol acetate**), based on a tigliane skeleton, found in the latex of many species in the Euphorbiaceae family, which are severe skin irritants and tumour promoters (Goel *et al.*, 2007). Plants containing these esters include *Croton* spp. and castor oil plant (*Ricinis communis*, Euphoriaceae) – care should be taken when handling them.

Table 5.3. Diterpenes in some major medicinal plants

Species	Compounds	Diterpenes class	Actions
Salvia rosmarinus (syn. *Rosmarinus officinalis*)	Carnosic acid; carnosol	Abietane	Antimicrobial, antioxidant
Stevia rebaudiana	Stevioside, rebaudioside	Ent-kaurane	Sweetener, antiviral
Andrographis paniculata	Andrographolide	Labdane	Hepatoprotective, immunostimulant
Plectranthus forskohlii (syn. *Coleus forskohlii*)	Forskolin	Labdane	Hypotensive, cardioprotective
Ginkgo biloba	Ginkgolides	Abietane	PAF inhibitor, neuroprotective
Taxus brevifolia, Taxus baccata	Taxol (paclitaxel)	Taxadiene	Antimitotic, anticancer
Marrubium vulgare	Marrubiin	Labdane	Expectorant
Dodonaea viscosa	Hautriwaic acid	Clerodane	Anti-inflammatory
Vitex agnus-castus	Vitexilactone	Labdane	Dopinomergic
Salvia divinorum	Salvinorin A	Neoclerodane	Halluginogenic

PAF, platelet activation factor.

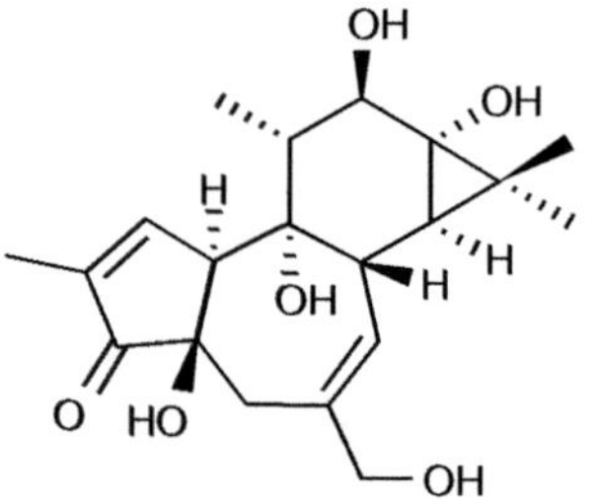

myristoylphorbol acetate – a diterpene phorbol ester

Teucrin A, an ent-clerodane diterpene, is a hepatotoxic compound found in wall germander (*Teucrium chamaedrys*, Lamiaceae). Concerns regarding hepatoxicity in skullcap herb (*Scutellaria lateriflora*, Lamiaceae) have been attributed to adulteration with wall germander (Gaffner and Blumenthal, 2016). **Andromedotoxin** is a neurotoxic diterpenoid found in various members of the Ericaceae family, including *Rhododendron* spp. The toxin is transferred to the nectar produced by these plants, ingestion of which can cause bradycardia, hypotension and risk of death (van Wyk and Wink, 2014).

Bitter principles

The term 'bitter principle' is used in reference to any one of a group of unrelated constituents responsible for the bitter taste characteristic of many herbs. Most are derived from terpenes, though non-terpene structures such as certain flavonoids and alkaloids are included.

Bitters have certain physiological effects regardless of their chemical structure, since the bitterness itself directly stimulates taste receptors, in turn sending signals via the gustatory nerve for the release of a cascade of gastric secretions and hormones. The effects are to stimulate appetite and digestive processes generally, increasing bile flow, regulating blood sugar levels and counteracting food sensitivities and allergies (Mills, 1997). The distribution of bitter receptors extends well beyond the oral cavity to the gastrointestinal and respiratory tracts (Rozengurt, 2006; Grassin-Delyle *et al.*, 2013), thereby extending the direct physiological effects of bitters to digestive and respiratory disorders.

Bitters are regarded as cooling remedies, hence beneficial for treating fevers and inflammation. In traditional Chinese medicine bitter herbs such as *Coptis chinensis* (Ranunculaceae) are used for treatment of diabetes (Chen *et al.*, 2015), while biomedical research has corroborated the role of bitters in diabetic treatments (Pham *et al.*, 2016).

Taste receptors are genetically linked to loci that influence bitter perceptions – hence individuals have vastly different abilities to perceive bitter tastes (Drewnowski and Gomez-Carneros, 2000). This genetic variability (polymorphism) has been interpreted as a protective mechanism for identification

Table 5.4. Bitter principles in herbs, with bitterness indices, where reported

Name	Class of compounds	Main constituents	Bitterness index
Centaurium spp.	Secoiridoid glycosides	Swertiamarin, centiopicroside	Plant > 2,000 Flowers 12,000
Gentiana lutea	Secoiridoid glycosides	Gentiopicrin 2–3.5% Amarogentin 0.5%	12,000 58×10^6 Root 10–30,000
Harpagophytum procumbens	Secoiridoid glycosides	Harpagoside, procumbide	Root 600–2,000
Olea europaea	Secoiridoid glycosides	Oleuropein 2%	NA
Menyanthes trifoliata	Monoterpene alkaloids	Foliamenthin, gentianin	Leaf 4–10,000
Plantago lanceolata	Iridoids	Aucubin 2–2.4%	NA
Cnicus benedictus	SLs	Cnicin, artemisiifolin	Herb 800–1,800
Artemisia absinthium	SLs	Absinthin 0.2%	12,700,000 Herb 10–25,000
Cynara scolymus	SLs	Cyanaropicrin	NA
Marrubium vulgare	Diterpenes	Marrubiin 0.3–1%	65,000
Bryonia alba	Triterpenes	Curcurbitacins	NA
Quassia arama	Secotriterpenes	Quassin 0.1–0.15%	17×10^6 Wood 40–50,000
Citrus × aurantium	Flavanone glycosides	Neohespiridin, naringin	500,000 Peel 600–1,500
Humulus lupulus	Phloroglucinols	Iso-humulone, lupulone	NA

NA, not available; SLs, sesquiterpene lactones.

of bitter poisons; however, it can also lead to rejection of healthy foods such as grapefruit and salad greens by many people, and the familiar problem of non-compliance with herbal medicines (Mennella *et al.*, 2013).

Table 5.4 is a list of herbal bitter constituents along with known bitterness indices, calculated from the maximum dilutions of constituents at which bitterness is still detectable (Wagner *et al.*, 1984).

Triterpenes

Triterpenes represent a very large group of medicinal plant compounds, and so a separate chapter has been devoted to them (Chapter 6, this volume).

Tetraterpenes

The main group of interest is the carotenoids, over 1000 of which are found in nature – providing red, orange and yellow pigments to fruits and vegetables.

Some, in particular α- and γ-**carotene**, act as provitamins, converted to vitamin A by the human digestive system.

Carotenoids are chemically characterized by long hydrocarbon chains with at least seven conjugated double bonds. They are synthesized from **phytoene**, a colourless C_{40} intermediate formed by condensation of two molecules of geranyl pyrophosphate. Phytoene is desaturated to form the yellow pigment **β-carotene**, with further enzymic reactions leading to the deep-red pigment lycopene (Maoka, 2019). **Lycopene**, found in tomatoes and other pink-red fruits and vegetables, has a simple acyclic structure, considered the prototype molecule for the carotenoid family.

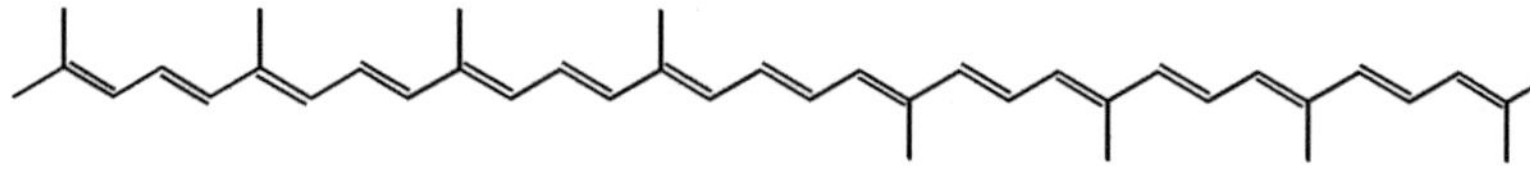

lycopene – an acyclic carotenoid

Lycopene is one of the most widely consumed carotenoids; however, it lacks provitamin A activity. Through cyclization at one or both end groups of lycopene, carotenes are formed, containing one (δ- and γ-carotenes) or two (α- and β-carotenes) ionone rings. **Vitamin A** (all-*trans*-retinol) consists of a β-ionone ring with a side chain of three isoprenoid units. Hence α- and β-carotenes provide two molecules of retinol in the human body whereas δ- and γ-carotenes provide only one. In reality, carotenoids are incompletely absorbed and 6 μg of β-carotene is equivalent to 1 μg retinol equivalent (RE). Carotenoids with hydroxylated ionone rings (e.g. lutein) provide no vitamin A activity (Eitenmiller and Landen, 1999).

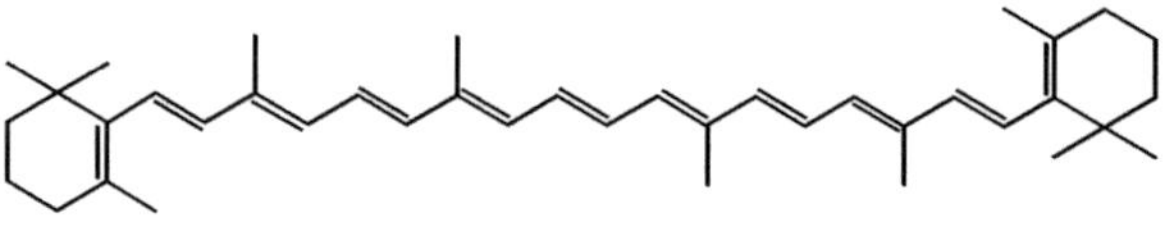

β-carotene

Despite having no provitamin A function, carotenoids such as lycopene and lutein are still biologically active. High serum lycopene concentrations have been linked to a reduced risk of atherosclerosis, while preventative effects against heart disease and cancer have also been reported. As an antioxidant it is a powerful quencher of oxygen free radicals compared to other carotenoids (Bruno and Wildman, 2001). Lycopene appears to play a particular role in reducing the risk of prostate cancer (Criston *et al.*, 2000).

References

Boatto, G., Pintore, G., Palomba, M., De Simone, F., Ramunda, E. and Jodice, C. (1994) Composition and antibacterial activity of *Inula helenium* and *Rosmarinus officinalis* essential oils. *Fitoterapia* 64, 279–280.

Bos, R., Woerdenbag, H.J. and Pras, N. (2002) Determination of valepotriates. *Journal of Chromatography A* 967, 131–146. https://doi.org/10.1016/S0021-9673(02)00036-5

Bruneton, J. (1995) *Pharmacognosy, Phytochemistry, Medicinal Plants.* Lavoisier Pubs, Paris.

Bruno, R.S. and Wildman, R.E.C. (2001) Sources, properties and nutraceutical potential of lycopene. *Journal of Nutraceuticals, Functional & Medical Foods* 3, 9–23. https://doi.org/10.1300/J133v03n02_02

Chadwick, M., Trewin, H., Gawthrop, F. and Wagstaff, C. (2013) Sesquiterpenoids lactones: benefits to plants and people. *International Journal of Molecular Science* 14, 12780–12805. doi: 10.3390/ijms140612780

Chander, R., Srivastava, V., Tandou, J.S. and Kapoor, N.K. (1995) Antihepatoxic activity of diterpenes of *Andrographis paniculata* (Kal-Megh) against *Plasmodium berghei*-induced hepatic damage in *Mastomys nututensis*. *International Journal of Pharmacognosy* 33, 135–138. https://doi.org/10.1016/0006-2952(93)90364-3

Chen, F., Tholl, D., Bohlmann, J. and Pichersky, E. (2011) The family of terpene synthases in plants: a mid-size family of genes for specialized metabolism that is highly diversified throughout the kingdom. *The Plant Journal* 66, 212–229. https://doi.org/10.1111/j.1365-313X.2011.04520.x

Chen, H., Guo, J., Pang, B., Zhao, L. and Tong, X. (2015) Application of herbal medicines with bitter flavor and cold property on treating diabetes mellitus. *Evidence-based Complementary and Alternative Medicine, eCAM* 2015, 529491. https://doi.org/10.1155/2015/529491

Cheng, A.-X., Lou, Y.-G., Mao, Y.-B., Lu, S., Wang, L.-J. and Chen, X.-Y. (2007) Plant terpenoids: biosynthesis and ecological functions. *Journal of Integrative Plant Biology* 49(2), 179–186. https://doi.org/10.1111/j.1744-7909.2007.00395.x

Criston, A., Di Pierro, F. and Bombardelli, E. (2000) Botanical derivatives for the prostate. *Fitoterapia* 71, S21–S28.

Cronquist, A. (1988) *The Evolution and Classification of Flowering Plants*, 2nd edn. New York Botanical Gardens, New York.

Das, B., Chowdhury, C., Kumar, D., Sen, R., Roy, R., *et al.* (2010) Synthesis, cytotoxicity, and structure–activity relationship (SAR) studies of andrographolide analogues as anti-cancer agents. *Bioorganic and Medicinal Chemistry Letters* 20, 6947–6950. doi: 10.1016/j.bmcl.2010.09.126

Dewick, P.M. (2009) The mevalonate and methylerythritol phosphate pathways: terpenoids and steroids. In: Dewick, P.M. (ed.) *Medicinal Natural Products: a Biosynthetic Approach*, 3rd edn. Wiley, Chichester, UK, pp. 187–310. doi: 10.1002/9780470742761

Dolara, P., Corte, B., Ghelardini, C., Pugliese, A.M., Cerbai, E., *et al.* (2000) Local anaesthetic, antibacterial and antifungal properties from myrrh. *Planta Medica* 66, 356–358. doi: 10.1055/s-2000-8532

Dondorp, A.M., Nosten, F., Yi, P., Das, D., Phyo, A.P., *et al.* (2009) Artemisinin resistance in *Plasmodium falciparum* malaria. *New England Journal of Medicine* 361(5), 455–467. doi: 10.1056/NEJMoa0808859

Drewnowski, A. and Gomez-Carneros, C. (2000) Bitter taste, phytonutrients and the consumer: a review. *American Journal of Clinical Nutrition* 72, 1424–1435. doi: 10.1093/ajcn/72.6.1424

Eitenmiller, R.E. and Landen Jr, W.O. (1999) *Vitamin Analysis for the Health and Food Sciences.* CRC Press, Boca Raton, Florida.

Fairhurst, R.M., Nayyar, G.M.L., Breman, J.G., Hallett, R., Vennerstrom, J.L. and Dondorp, A.M. (2012) Artemisinin-resistant malaria: research challenges, opportunities, and public health implications. *The American Journal of Tropical Medicine and Hygiene* 87(2), 231–241. https://doi.org/10.4269/ajtmh.2012.12-0025

Farooqi, A.A., Fayyaz, S., Silva, A.S., Sureda, A., Nabavi, S.F., *et al.* (2017) Oleuropein and cancer chemoprevention, the link is hot. *Molecules* 22(5), 705. doi: 10.3390/molecules22050705

Ferriera, E.B., Neves, F.R., Da Costa, M.A.D., Do Prado, W.A., Ferri, L.F. and Bazotte, R.B. (2006) Comparative effects of *Stevia rebaudiana* leaves and stevioside on glycaemia and hepatic glucogenesis. *Planta Medica* 72(8), 691–696. doi: 10.1055/s-2006-931586

Ferreira, J.F.S., Peaden, P. and Keiser, J. (2011) *In vitro* trematocidal effects of crude alcoholic extracts of *Artemisia annua, A. absinthium, Asimina triloba,* and *Fumaria officinalis. Parasitology Research* 109(6), 1585–1592. doi: 10.1007/s00436-011-2418-0

Fraga, B.M. (2012) Natural sesquiterpenoids. *Natural Product Reports* 29(11), 1334–1336. doi: 10.1039/c2np20074k

Gaffner, S. and Blumenthal, M. (2016) On adulteration of skullcap. *Botanical Adulterants Bulletin* March, 1–5. Available at: http://cms.herbalgram.org/BAP/BAB/BAP-BABs-Skullcap-CC-FINAL.pdf (accessed 30 November 2020).

Goel, G., Makkar, H.P.S., Francis, G. and Becker, K. (2007) Phorbol esters: structure, biological activity, and toxicity in animals. *International Journal of Toxicology* 26(4), 279–288. doi: 10.1080/10915810701464641

Grassin-Delyle, S., Abrial, C., Fayad-Kobeissi, S., Brollo, M., Faisy, C., *et al.* (2013) The expression and relaxant effect of bitter taste receptors in human bronchi. *Respiratory Research* 14, 134. Available at: http://respiratory-research.com/content/14/1/134 (accessed 30 November 2020).

Guba, R. (2018) The modern alchemy of carbon dioxide extraction. *Aromatherapy Today* (special edition) August, 35–39.

Hamburger, M. and Hostettmann, K. (1991) Bioactivity in plants: the link between phytochemistry and medicine. *Phytochemistry* 30(12), 3864–3874. https://doi.org/10.1016/0031-9422(91)83425-K

Harborne, J. and Baxter, H. (1993) *Phytochemical Dictionary.* Taylor & Francis, London.

Hassan, Z.K., Elamin, M.H., Daghestani, M.H., Omer, S.R., Al-Olayan, E.M., *et al.* (2012) Oleuropein induces anti-metastatic effects in breast cancer. *Asian Pacific Journal of Cancer Prevention* 13(9), 4555–4559. doi: 10.7314/apjcp.2012.13.9.4555

Hormann, H. and Korting, H. (1995) Allergie acute dermatitis due to Arnica tincture self-medication. *Phytomedicine* 1(4), 315–317. doi: 10.1016/S0944-7113(11)80009-7

Huang, K. (1993) *The Pharmacology of Chinese Herbs.* CRC Press, Boca Raton, Florida.

Huang, S.H., Duke, R.K., Chebib, M., Sasaki, K., Wada, K. and Johnston, G.A.R. (2004) Ginkgolides, diterpene trilactones of *Ginkgo biloba,* as antagonists at recombinant $\alpha 1\beta 2\gamma 2L$ GABA$_A$ receptors. *European Journal of Pharmacology* 494, 131–138.

Jiang, H., Chen, J., Jin, X., Yang, J., Li, Y., *et al.* (2011) Sesquiterpenoids, alantolactone analogues, and seco-guaiene from the roots of *Inula helenium. Tetrahedron* 67(47), 9193–9198. doi: 10.1016/j.tet.2011.09.070

Klaas, C.A., Wagner, G., Laufer, S., Sosa, S., Della Loggia, R., *et al.* (2001) Studies on the anti-inflammatory activity of phytopharmaceuticals prepared from Arnica flowers. *Planta Medica* 68(5), 385–391. doi: 10.1055/s-2002-32067

Kuzuyama, T. and Seto, H. (2012) Two distinct pathways for essential metabolic precursors for isoprenoid biosynthesis. *Proceedings of the Japan Academy. Series B, Physical and Biological Sciences* 88(3), 41–52. https://doi.org/10.2183/pjab.88.41

Lee, B., Shin, Y.W., Bae, E.A., Han, S.J., Kim, J.S., *et al.* (2008) Antiallergic effect of the root of *Paeonia lactiflora* and its constituents paeoniflorin and paeonol. *Archives of Pharmaceutical Research* 31(4), 445–450. doi: 10.1007/s12272-001-1177-6

Lenaz, L. and De Furia, M.D. (1993) Taxol: a novel natural product with significant anticancer activity. *Fitoterapia LXIV*, 27–36.

Liu, Y.-F., Liang, D., Luo, H., Hao, Z.-Y., Wang, Y., *et al.* (2012) Hepatoprotective iridoid glycosides from the roots of *Rehmannia glutinosa*. *Journal of Natural Products* 75(9), 1625–1631. https://doi.org/10.1021/np300509z

Madan, S., Ahmad, S., Singh, G.N., Kohli, K., Kumar, Y., *et al.* (2010) *Stevia rebaudiana (Bert.) Bertoni* – a review. *Indian Journal of Natural Products and Resources* 1(3), 267–286. Available at: http://nopr.niscair.res.in/handle/123456789/10287 (accessed 30 November 2020).

Majdi, M., Liu, Q., Karimzadch, G., Malboobi, M.A., Beekwilder, J., *et al.* (2011) Biosynthesis and localization of parthenolide in glandular trichomes of feverfew (*Tanacetum parthenium* L. Schulz Bip.). *Phytochemistry* 72(14–15), 1739–1750. doi: 10.1016/j.phytochem.2011.04.021

Maoka, T. (2019) Carotenoids as natural functional pigments. *Journal of Natural Medicines* 74, 1–16. https://doi.org/10.1007/s11418-019-01364-x

Mata, R., Rivero-Cruz, J.F. and Rojas, A. (1999) Smooth muscle-relaxing constituents of *Dodonaea viscosa* (Sapindaceae) and *Conyza filaginoides* (Asteraceae). In: Hostettmann, K., Gupta, M.P. and Marston, A. (eds) *Proceedings of the IOCD/CYTED Symposium, 23–26 February, 1997, Panama City, Panama*. Harwood Academic Publishers, Amsterdam, pp. 161–183.

Melchior, J., Spasov, A.A., Ostrovskij, O.V., Bulanov, A.E. and Wikman, G. (2000) Double-blind, placebo cross-over pilot and Phase III study of standardised *Andrographis paniculata* herba Nees extract fixed combination in the treatment of uncomplicated upper respiratory tract infection. *Phytomedicine* 7(5), 341–350. doi: 10.1016/S0944-7113(00)80053-7

Menendez, J.A., Vazquez-Martin, A., Garcia-Villalba, R., Carrasco-Pancorbo, A., Oliveras-Ferraros, C., *et al.* (2008) tabAnti-HER2 (*erb*B-2) oncogene effects of phenolic compounds directly isolated from commercial extra-virgin olive oil (EVOO). *BMC Cancer* 8, 377. https://doi.org/10.1186/1471-2407-8-377

Mennella, J.A., Spector, A.C., Reed, D.R. and Coldwell, S.E. (2013) The bad taste of medicines: overview of basic research on bitter taste. *Clinical Therapeutics* 35(8), 1225–1246. doi: 10.1016/j.clinthera.2013.06.0

Mills, S. (1997) Acupharmacology, Pt. 2. *Modern Phytotherapist* 3(3), 12–15.

Omar, S.H. (2010) Oleuropein in olive and its pharmacological effects. *Scientia Pharmaceutica* 78(2), 133–154. https://doi.org/10.3797/scipharm.0912-18

Padovan, A., Keszei, A., Külheim, C. and Foley, W.J. (2014) The evolution of foliar terpene diversity in Myrtaceae. *Phytochemistry Review* 13, 695–716. doi: 10.1007/s11101-013-9331-3

Pham, H., Hui, H., Morvaridi, S., Cai, J., Zhang, S., *et al.* (2016) A bitter pill for type 2 diabetes? The activation of bitter taste receptor TAS2R38 can stimulate GLP-1 release from enteroendocrine L-cells. *Biochemical Biophysical Research Communications* 475(3), 295–300. doi: 10.1016/j.bbrc.2016.04.149

Qi, J., Li, N., Zhou, J., Yu, B. and Qiu, S.X. (2010) Isolation and anti-inflammatory activity evaluation of triterpenoids and a monoterpenoid glycoside from *Harpagophytum procumbens*. *Planta Medica* 76(16), 1892–1896. doi: 10.1055/s-0030-1250029

Rateb, M.E., El-Gendy, A.A.M., El-Hawary, S.S. and El-Shamy, A.A.M. (2007) Phytochemical and biological investigation of *Tanacetum parthenium* (L.) cultivated in Egypt. *Journal of Medicinal Plant Research* 1(1), 018–026. doi: 10.5897/JMPR

Recio, M., Giner, R., Manez, S. and Rios, J. (1994) Structural considerations on the iridoids as anti-inflammatory agents. *Planta Medica* 60(3), 232–234. doi: 10.1055/s-2006-959465

Rodriguez, E., Towers, G.H.N. and Mitchell, J.C. (1976) Biological activities of sesquiterpene lactones. *Phytochemistry* 15(11), 1573–1580. https://doi.org/10.1016/S0031-9422(00)97430-2

Rodriguez, S., Marston, A., Wolfender, J.-L. and Hostettmann, K. (1998) Iridoids and secoiridoids in the Gentianaceae. *Current Organic Chemistry* 2(6), 627–648.

Rozengurt, E. (2006) Taste receptors in the gastrointestinal tract. *American Journal of Physiology –Gastrointestinal and Liver Physiology* 291, G171–G177. doi: 10.1152/ajpgi.00073.2006

Salinas-Sánchez, D.L., Herrera-Ruiz, M., Pérez, S., Jiménez-Ferrer, E. and Zamilpa, A. (2012) Anti-inflammatory activity of hautriwaic acid isolated from *Dodonaea viscosa* leaves. *Molecules* 17, 4292–4299. doi: 10.3390/molecules17044292

Seo, J.Y., Park, J., Kim, H.J., Lee, I.A., Lim, J.-S., *et al.* (2009) Isoalantolactone from *Inula helenium* caused Nrf2-mediated induction of detoxifying enzymes. *Journal of Medicinal Food* 12(5), 1038–1045. doi: 10.1089/jmf.2009.0072.

Spelman, K. (2009) 'Silver bullet' drugs vs. traditional herbal remedies: perspectives on malaria. *HerbalGram* 84, 44–55.

Sticher, O. (1977) Plant mono-, di- and sesquiterpenoids with pharmacological or therapeutic activity. In: Wagner, H. and Wolff, P. (eds) *New Natural Products with Pharmacological, Biological or Therapeutic Activity*. Springer, Berlin, pp. 137–176.

Tada, M., Okuno, K., Chiba, K., Ohnishi, E. and Yoshii, T. (1994) Antiviral diterpenes from *Salvia officinalis*. *Phytochemistry* 35(2), 539–541. https://doi.org/10.1016/S0031-9422(00)94798-8

Tisserand, R. and Balacs, T. (1995) *Essential Oil Safety*. Churchill Livingstone, Edinburgh, UK.

Van Wyk, B. (2013) *Culinary Herbs and Spices of the World*. Kew Publishing, Richmond, UK.

Van Wyk, B. and Wink, M. (2014) *Phytomedicines, Herbal Drugs and Poisons*. Kew Publishing, Richmond, UK.

Visen, P.K.S., Shukla, B., Patnaik, G.K. and Dhawan, B.N. (1993) Prevention of galactosamine-induced hepatic damage by picroliv. *Planta Medica* 59, 37–41.

Vissers, M.N., Zock, P.L., Roodenburg, A.J.C., Leenen, R. and Katan, M.B. (2002) Olive oil phenols are absorbed in humans. *Journal of Nutrition* 132(3), 409–417. Available at: https://www.ncbi.nlm.nih.gov/pubmed/11880564 (accessed 30 November 2020).

Wagner, H., Bladt, S. and Zgainski, E. (1984) *Plant Drug Analysis*. Springer, Berlin.

Triterpenoids and Saponins

Introduction

Triterpenoid compounds are derived from a C_{30} precursor, squalene, which was first isolated from shark liver, and is also found in yeast and some seed oils. Squalene comes about by tail-to-tail coupling of farnesyl pyrophosphate (FPP), a product of the mevalonate pathway (Dewick, 2009). Cyclization of squalene occurs by the opening of the epoxide in 2,3-oxidosqualene, leading to formation of cyclohexane and cyclopropane rings, both of which can occur in both *cis* and *trans* forms. Further biosynthesis involving 2,3-oxidosqualene cyclase enzyme leads to formation of the more than 4000 known triterpenoid compounds and 400 structures isolated from plants (Kreis and Müller-Uri, 2010). Most are hydroxylated at C-3. Triterpenoids are lipophilic (soluble in fats, ethanol, insoluble in water), except when bound to sugars in glycoside form, for example as saponins.

2,3-oxidosqualene

squalene   β–amyrin

triterpenoids (e.g. β-amyrin) are derived from squalene, in a process mediated by the enzyme 2,3-oxidosqualene cyclase

Steroids, found in both plants and animals, have similar triterpene configuration, their reduced C_{27} skeletons are also derived from squalene. According to convention, the rings of triterpenes are labelled A–D (or E if present) and the carbons are numbered as shown on the diagram.

DOI: 10.1079/9781789243079.0006   95

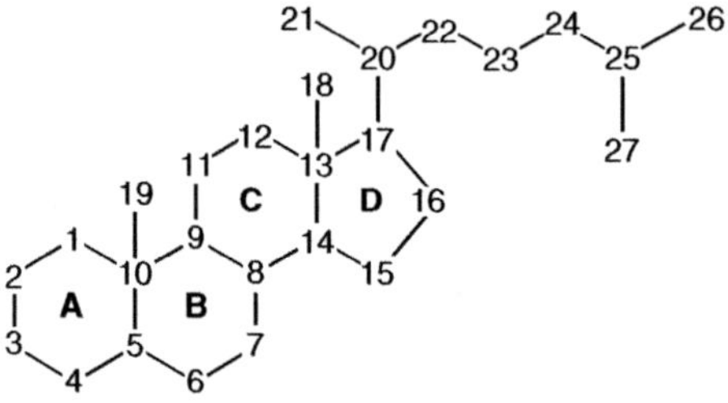

triterpenoid numbering convention

The triterpenoids are a large and diverse group made up of several sub-classes. These include:

- free triterpenes;
- triterpenoid saponins;
- steroidal saponins;
- cardiac glycosides;
- phytosterols;
- phytoecdysteroids;
- curcurbitacins; and
- quassinoids.

These may occur in the free state within plants, or as aglycones of glycosides. The major classes of triterpenoids are shown in Table 6.1.

Free triterpenoids

Calendula officinalis (Asteraceae) is a rich source of both free triterpenoids and triterpenoid glycosides (Muley *et al.*, 2009). In a bioassay-oriented fractionation of *C. officinalis* flower extracts, the most potent anti-inflammatory activity was found to lie in the free triterpenes, rather than their glycosides. The most active of these was faradiol monoester (Della Loggia *et al.*, 1994). The presence of faradiol monoester is regarded as a good index of anti-inflammatory activity (Szakiel *et al.*, 2005).

Non-glycosidic triterpenes, usually of the lanostane class, are found in large concentrations in some of the medicinal macrofungi – in particular, reishi (*Ganoderma lucidum*, Gandodermaceae) and *Poria cocos* (Polyporaceae). The most significant are highly oxygenated ganoderals, ganoderic and lucideric acids from reishi. More than 130 triterpenes have been isolated from the fruit and mycelium of *G. lucidum* (Gao *et al.*, 2002). Triterpenes are responsible for the bitter taste that distinguishes reishi from many other mushrooms. These compounds inhibit histamine release, are antihepatotoxic and angiotensin-converting enzyme (ACE)-inhibiting. They also inhibit cholesterol synthesis (Komoda *et al.*, 1989) and some were proved to have cytotoxicity against hepatoma cells *in vitro* (Hirotani *et al.*, 1985). Ten lanostane triterpenes have

Table 6.1. Major classes of triterpenoids with examples

Triterpene class	Parent compound	Structure	Medicinal compounds	Plant source
Oleanane	β-Amyrin	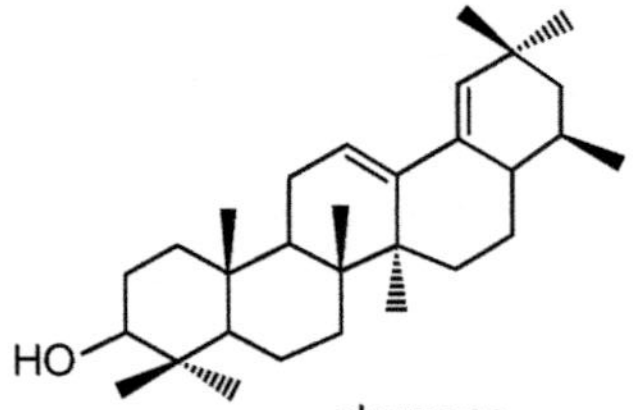 oleanane	Glycyrrhizin α-Hederin Saikosaponin Arvenoside Soyasaponin	*Glycyrrhiza glabra* *Hedera helix* *Bupleurum falcatum* *Calendula officinalis* *Glycine max*
Dammarane	Protopanaxadiol	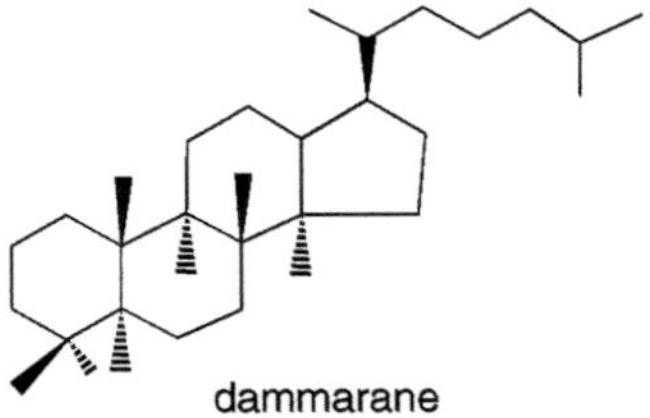 dammarane	Ginsenosides Rg1, Rb1 Bacopasaponins Jujubosides	*Panax ginseng* *Bacopa monnieri* *Ziziphus jujuba*
Ursane	α-Amyrin	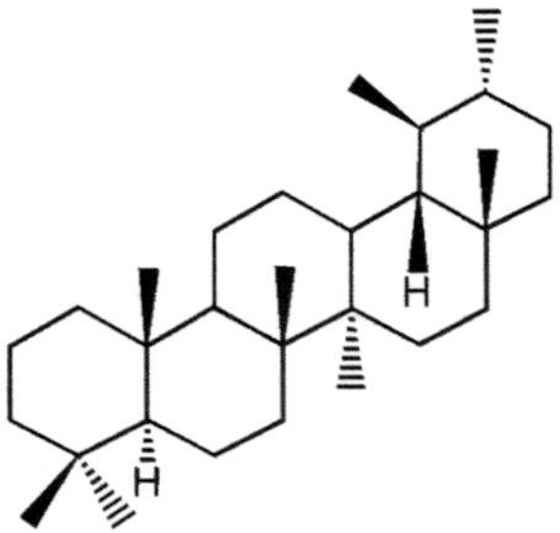 ursane	Asiatoside Asiatic acid Ursolic acid Faradiol monoester	*Centella asiatica* *Eucalyptus camaldulensis* *Rosmarinus officinalis* *Calendula officinalis*

Continued

Table 6.1. Continued.

Triterpene class	Parent compound	Structure	Medicinal compounds	Plant source
Lanostane	Lanosterol	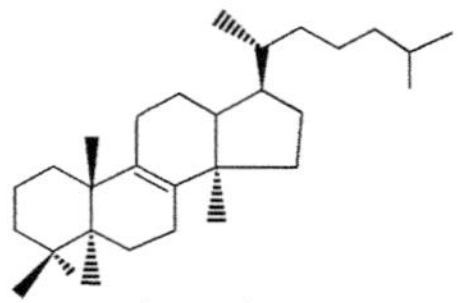 lanostane	Ganosterols, ganoderic acids Coriacoic acid	*Ganoderma lucidum* *Poria cocos*
Lupane	Lupeol	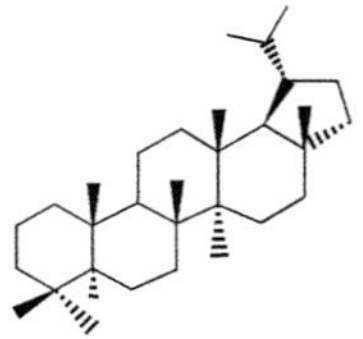 lupane	Betulin, betulinic acid Officinatrione, faradiol	*Betula* spp. *Taraxacum officinale*
Cycloartane	Cycloartenol	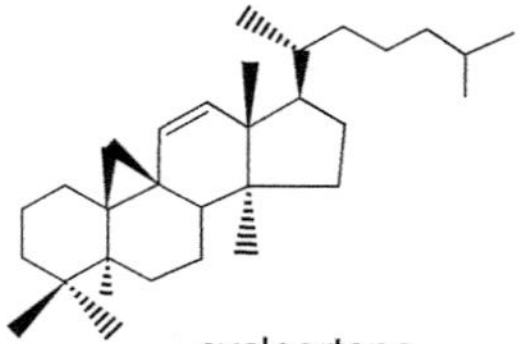 cycloartane	23-Epi-26-deoxyactein Actein, astragaloside IV	*Actaea racemosa* *Astragalus mongholicus*
Spirostane	Cholesterol	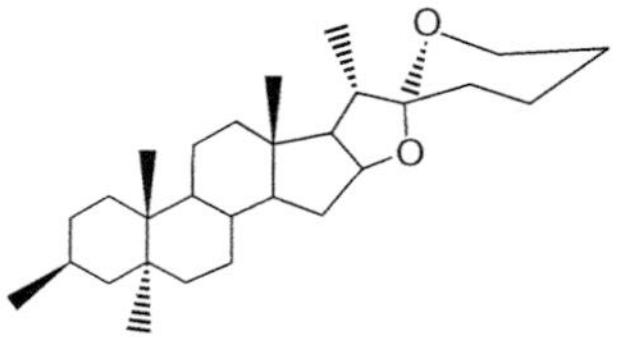 spirostane	Diosgenin Sarsasapogenin Ruscogenins	*Dioscorea villosa* *Smilax* spp. *Tribulus terrestris* *Ruscus aculeatus*

been found to be ACE inhibitors – these compounds were responsible for *in vivo* antihypertensive activity in mice (Morigiwa *et al.*, 1986). Several of the terpenes demonstrate antiviral action against HIV-1, influenza A virus and herpes simplex virus type 1 (HSV-1) (Liu *et al.*, 2005; Zhu *et al.*, 2015).

Poriatin from *P. cocos* has immunomodulating properties. It has demonstrated antiviral activity and increased production of macrophages and other immune cells. It has also demonstrated immunosuppressive activity, reducing the severity of induced autoimmune encephalitis in conjunction with a standard autoimmune drug. Poriatin is a known aldosterone antagonist (Hobbs, 1995).

In a recent review specific anticancer activity has been reported for triterpenoids of the lupane, oleanane, ursane, cucurbitane and dammarane classes, via multiple mechanisms such as modulation of inflammation-associated signalling pathways including NF-κB (nuclear factor kappa B), and downregulation of the expression of protein transcription factors in cancer cells (Ren and Kinghorn, 2019).

Saponins

Saponins are compounds whose active portions form colloidal solutions in water, which produce lather on shaking and precipitate cholesterol. They occur as glycosides whose aglycones are triterpenoid or steroidal structures. The combination of lipophilic (fat-soluble) aglycones at one end of the molecule, and hydrophilic (water-soluble) sugars at the other gives them the ability to lower surface tension, producing the characteristic detergent or soap-like effect on membranes and skin. This action is responsible for the effectiveness of saponin-rich plants as traditional fish poisons, and for numerous industrial applications of isolated saponins such as in foaming agents, cosmetics and pharmaceutical products (Guclu-Ustundag and Massa, 2007; Vincken *et al.*, 2007). Plant-based soaps are mostly derived from saponin-rich plants such as soapwort (*Saponaria officinalis*, Caryophyllaceae) (Sawai and Saito, 2011).

Triterpenoid saponins

The most widely distributed triterpenoid aglycone is oleanolic acid, which forms a pentacyclic structure referred to as the oleanane-type ring system. Glycyrrhizic acid from liquorice root (*Glycyrrhiza glabra* and *Glycyrrhiza uralensis*, Fabaceae) is an example. In this case the sugar component consists of two units of glucuronide, metabolites of glucose. Glycyrrhizin is a mixture of calcium and potassium salts of glycyrrhizic acid (Dewick, 2009). Other triterpenoid ring systems include ursane and lupane types as well as the dammarane type – represented by ginsenosides from *Panax ginseng* (Araliaceae) and jujubosides from *Ziziphus jujuba* (Rhamnaceae) (see Table 6.1).

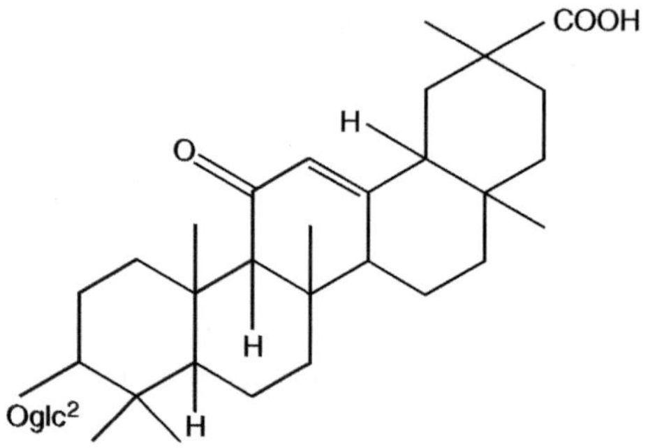

glycyrrhizic acid from *Glycyrrhiza glabra*

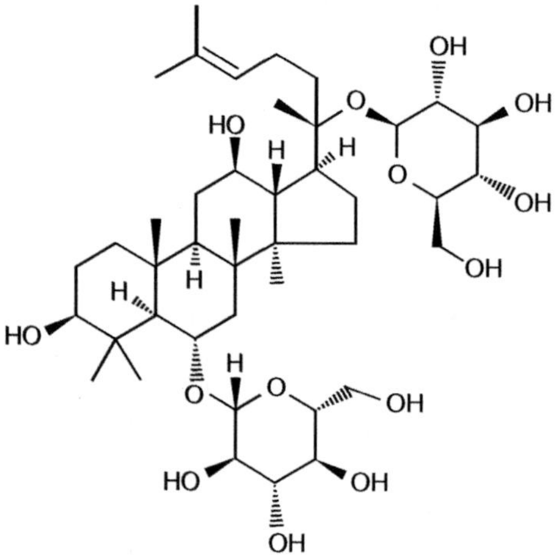

ginsenoside Rg1

Steroidal saponins

These are not true triterpenes since their C_{27} ring skeletons cannot be broken down into isoprene units, although they have a common biosynthetic origin to triterpenes via the mevalonic acid pathway. They are sometimes referred to as nortriterpenes. Some including diosgenin and hecogenin are used as precursors of sex hormones, cortisone and vitamin D. Steroidal saponins are thought to be responsible, at least, in part, for the oestrogenic activity linked to, for example, *Dioscorea villosa* (Dioscoreaceae) and *Chamaelirium luteum* (Melanthiaceae), though they are not usually classed among the phytoestrogens.

A subgroup of the steroidal saponins are the glycoalkaloids, in which the aglycone is a steroidal alkaloid (contains a nitrogen atom), for example solasodine (see diagram) which is poisonous. The most common source of these compounds is the *Solanum* genus, including the common potato (*Solanum tuberosum*, Solanaceae).

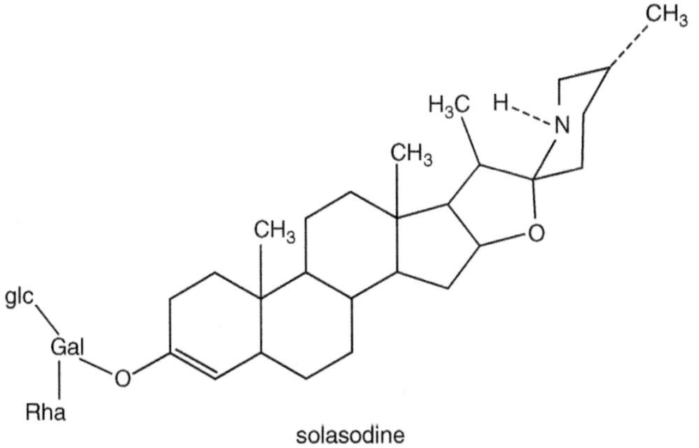

solasodine

Saponins can also be classified according to the way the aglycones are bonded to their sugar moieties:

- Monodesmosidic saponins have their sugars and aglycones linked by a single hydroxyl (OH) group.
- Bisdesmodic saponins are linked by two OH groups or one OH and one carboxyl group.

Pharmacological actions of saponins

Numerous saponin-rich herbs are known for their local detergent and wound-healing effects, these include *Calendula officinalis* referred to above, and gotu kola (*Centella asiatica*, Apiaceae). Gotu kola contains mixtures of saponins and free triterpenoids of either ursane or oleanane structural types (James and Dubery, 2009). Traditional and modern uses for gotu cola for topical applications include wound healing, eczema and psoriasis, burn and scar treatment, skin infections including leprosy and for revitalizing connective tissue (James and Dubery, 2009).

Saponins also have a range of systemic effects according to their chemical structures (see Table 6.2).

Many of the traditional herbal expectorants and diuretics contain significant amounts of saponins. Examples include senega (*Polygala senega*, Polygalaceae), golden rod (*Solidago virgaurea*, Asteraceae), ivy leaf (*Hedera helix*, Araliaceae) and *Primula* spp. (Primulaceae). In sufficient doses these saponins irritate the vagus nerve in the stomach, leading to production of increased secretions in the bronchial and renal tubes via reflex action. This secretolytic effect is responsible for their expectorant and diuretic properties.

Given their structural relationship to steroids, many saponins are known for their anti-inflammatory effects. Among the most potent are saikosaponins from *Bupleurum falcatum* (Apiaceae) and boswellic acid from *Boswellia serrata* (Burseraceae). Saikosaponins are also immunomodulatory and hepatoprotective (Yen *et al.*, 1994). Some saponins have been found to reduce inflammation

Table 6.2. Systemic effect of various saponins (from Lacaille-Dubois and Wagner, 1996; Guclu-Ustundag and Massa, 2007)

Anticancer: cytotoxic, antitumour, antimutagenic
Anti-inflammatory, antiallergic
Immunomodulatory, adjuvants for vaccines
Antimicrobial: antiviral, antifungal
Hepatoprotective
Antidiabetic: hypoglycaemic
Molluscicidal
Cardiac activity: haemolytic, antithrombotic, hypocholesterolaemic
Central nervous system activity: anti-stress, sedative
Endocrine activity: adaptogenic, oestrogenic
Expectorant, diuretic
Digestive – accelerate the body's ability to absorb nutrients

by inhibition of complement activity – these include some ginseng saponins, kaikasaponins and soyasaponins from kudzu (*Pueraria lobata*, Fabaceae) (Oh *et al.*, 2000). The presence of certain chemical groups within the saponin skeletons appears to influence anti-inflammatory activity, for example a carboxylic group at C-28 or C-30 (Recio *et al.*, 1995).

Saponins can also act as immunostimulants, acting to stimulate both the T helper (Th1) immune response and the production of cytotoxic T lymphocytes, providing a basis for their application in human and veterinary vaccines (Moses *et al.*, 2014). Astragaloside IV from the Chinese herb *Astragalus mongholicus* (syn. *A. membranaceus*, Fabaceae) is a potent immunostimulant. It achieves this effect by enhancing expression of major histocompatibility complex molecules on the surface of dendritic cells, as well as by modulation of various cytokines resulting in enhanced T-cell responses (Ren *et al.*, 2013). Astragaloside IV also reduced inflammation via inhibition of nitric oxide in macrophages (Lee *et al.*, 2013).

Saponins, along with phytosterols, are known to play a role in reducing cholesterol plasma concentrations by increasing permeability of the lipid membrane of cholesterol (Moses *et al.*, 2014). Diosgenin, the aglycone of dioscin in *Dioscorea villosa*, has been shown to reduce cholesterol absorption *in vivo*, leading to increased hepatic synthesis and excretion (Son *et al.*, 2007). Diosgenin reduced triglyceride levels and inhibited the expression of lipogenic genes in HepG2 cells, while also increasing the expression of genes involved in cholesterol synthesis (Patel *et al.*, 2012). Diosgenin is most well known as the starting material for the partial synthesis of steroid and contraceptive drugs (Hu *et al.*, 2007; Patel *et al.*, 2012).

Many vegetables, including spinach, tomatoes and asparagus as well as legumes (especially soybeans), are rich in saponins – it is quite possible they also help reduce plasma cholesterol in this way.

Saponins in herbs such as horse chestnut (*Aesculus hippocastanum*, Sapindaceae) and butcher's broom (*Ruscus aculeatus*, Asparagaceae) are of benefit to the smaller blood vessels. Their capillary protective and venous toni-fying effects lead to reduction of swelling and inflammation, they have dem-onstrated clinical benefits in the treatment of chronic venous insufficiency, and prevention of pain and cramps in the legs (Guclu-Ustundag and Massa, 2007; Urbanek, 2017). *Centella asiatica* (Apiaceae) also benefits microcirculatory dis-orders, notably in diabetic patients. In a randomized double placebo-controlled pilot study, a triterpenoid-rich extract of *C. asiatica* reduced the severity of symp-toms of diabetic neuropathy in humans with type 2 diabetes (Lou *et al.*, 2018).

Some Fabaceae saponins have liver-protecting effects. Both soyasaponins found in soybeans (*Glycine max*) and kudzusaponins found in kudzu root (*Pueraria* spp.) have been shown to be hepatoprotective in rodents (Arao *et al.*, 1998).

Numerous triterpenoid saponins have been shown to inhibit pathogenic fungi, for example monodesmosides α- and β-hederin from the common ivy (*Hedera helix*, Araliaceae). Antifungal activity appears to be influenced by the number and kinds of sugar residues (Favel *et al.*, 1994).

Safety issues relating to saponins

As noted above, saponins are capable of increasing permeability of mem-branes. They can cause haemolysis by destroying the membranes of red blood cells, thus releasing the haemoglobin. This doesn't occur when saponins are taken orally. However, if they are injected, this may lead to anaemia and renal failure, due to the sudden influx of haemoglobin into the bloodstream. This ef-fect has been observed in both *in vivo* and *in vitro* research experiments.

The tendency for some saponins to demonstrate greater haemolytic effects compared to others has been linked to the number of sugar linkages and the structure of the aglycone, including the point at which sugar linkages occur (Takechi and Tanaka, 1995; Dewick, 2009; Moses *et al.*, 2014).

When taken orally, saponins are absorbed rather poorly from the gut. Those that are absorbed are often in the aglycone form, following interaction with bacteria in the large bowel. The slow rate of absorption significantly slows the rate of haemolysis and associated toxicity. Saponins are regularly consumed in everyday foods such as cereals (Osbourn, 2003) and, with few exceptions, do not demonstrate toxic effects in humans (Guclu-Ustundag and Massa, 2007).

Phytosterols

Sterols such as **stigmasterol**, **sitosterol** and **campesterol** are essential com-ponents of plant cell membranes, and they are also used as the starting ma-terial in the production of steroidal drugs. Phytosterols are characterized by an OH group attached to a triterpene backbone at C-3 and an extra methyl or ethyl substituent in the side chain, the latter are not present in animal sterols (Harborne and Baxter, 1993). Sterols are found in a wide variety of cereal grains, pulses, seeds and to a lesser extent in vegetables. They also occur in the cell membranes of fungi and algae.

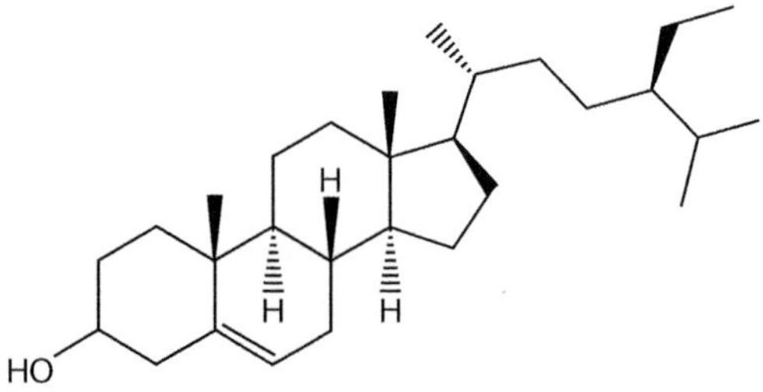

sitosterol

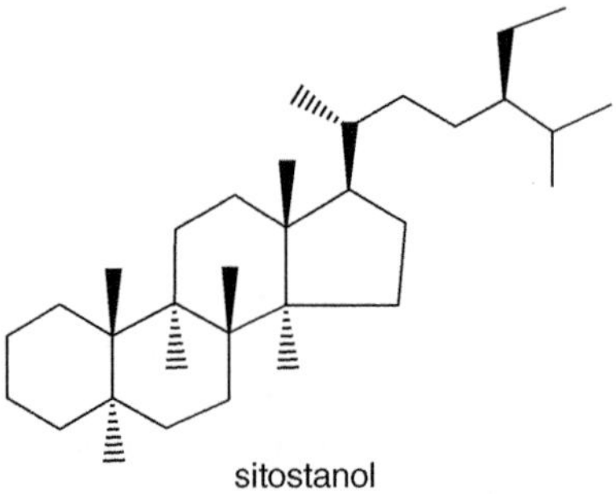

sitostanol

Phytosterols are minor but beneficial components of the human diet, helping in regulation of blood cholesterol. They may occur in the form of free alcohols, also as esters of fatty acids and as glycosides. Sterols may also occur in a fully saturated form, in which case they are known as phytostanols (e.g. **sitostanol** – see diagram). Functional foods in the form of yogurt and margarine, fortified with various forms of sterols, are widely marketed, and promoted for their ability to reduce low density lipoprotein (LDL) cholesterol levels (Lagarda *et al.*, 2004). Soy and rapeseed have been genetically modified (GM) to produce oils with increased stanol levels, as these are believed to be more potent hypocholesterolaemic agents (Dewick, 2009). Hence many of the fortified cholesterol-lowering functional foods may be produced from GM plants.

Phytosterols provide further benefits to the diet by acting as chemoprotectives against various forms of cancer. The only downside to their consumption is an associated reduction in absorption of α-tocopherols and β-carotenoids, sources of vitamins E and A, respectively (Woyengo *et al.*, 2009).

The resin extracted from myrrh (*Commiphora mukul*, Burseraceae) (known as guggul in India), contains sterols and related compounds known as guggulsterones. Guggul extracts are widely marketed in Asia for lowering blood cholesterol and triglycerides. Several hypolipidaemic mechanisms have been demonstrated, including stimulation of thyroid function. Guggul extracts provide various health benefits, in addition to being cardioprotective they have anticancer, anti-obesity, antidiabetic and neuroprotective effects (Shah *et al.*, 2012).

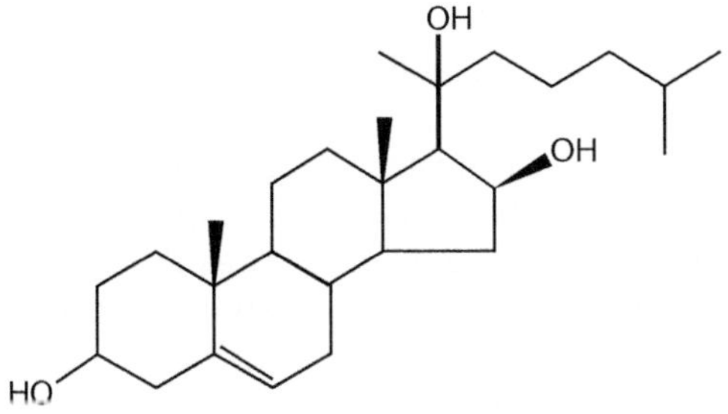

guggulsterol-II

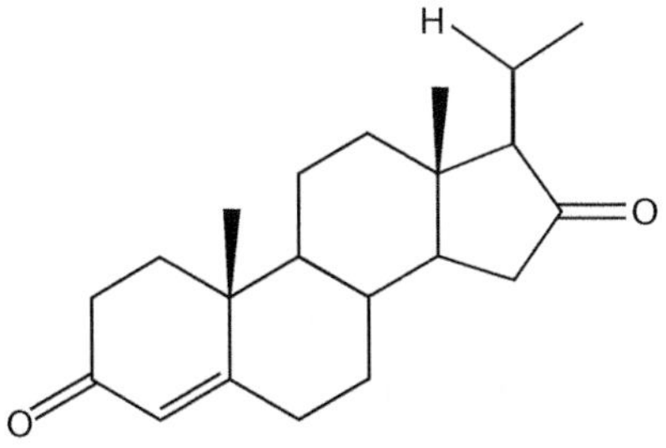

guggulsterone-Z

Investigations into stinging nettle root (*Urtica dioica*, Urticaceae) have demonstrated potent inhibition of enzymes involved in benign prostatic hyperplasia (BPH). Sterol compounds including **stigmast-4-en-3-one** and sitosterol are thought to be responsible for this BPH-inhibiting activity, while the sterols also inhibit aromatase, an enzyme involved in the genesis of prostate cancer (Hirano *et al.*, 1994; Asgarpanah and Mohajerani, 2012).

Herbal adaptogens

These agents are referred to as 'harmony remedies' by Stephen Fulder in his excellent book originally titled *The Root of Being: Ginseng and the Pharmacology of Harmony* (Fulder, 1980), still one of the best analyses of this class of medicinal agent. The concept of adaptogens is based around enhancement of vitality and general resistance rather than treatment of specific illnesses.

'By helping the body cope with stress, adaptogens can help accelerate learning speed, improve the memory, increase stamina in high performance athletes, alleviate small complaints and cut down infections by acting as a prophylactic' (Wahlstrom, 1987).

The contemporary authority on adaptogens, Panossian, includes terpenes with a tetracyclic skeleton among the main phytochemical groups responsible for adaptogenic activity (Panossian, 2017). These include ginsenosides, cucurbitacins, sitoindosides and withanolides – the latter two groups come from *Withania somnifera* (Solanaceae).

W. somnifera, known as ashwaganda in Ayurvedic medicine, contains steroidal lactones called **withanolides**, C_{28} compounds based on an ergostane skeleton. Withaferin A (see diagram) was the first withanolide to be characterized, but since then hundreds of others have been isolated (Chen *et al.*, 2010). While these constituents contribute to the adaptogenic action of ashwaganda referred to above, other actions based on research include anti-inflammatory, antitumour, immunomodulatory, antimicrobial, antiparasitic, hepatoprotective and antioxidant (Budhiraja *et al.*, 2000; Chen *et al.*, 2010).

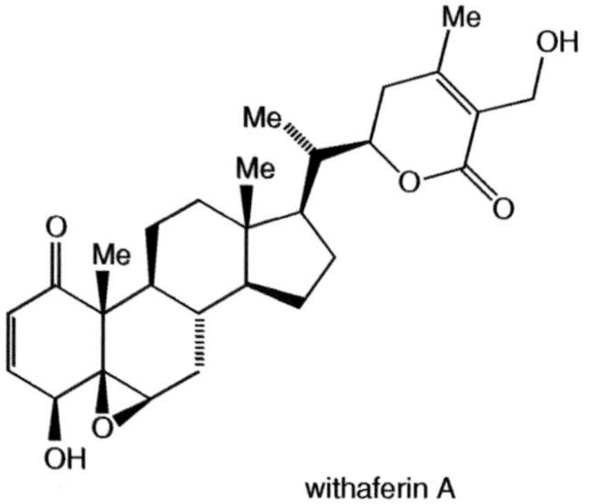

withaferin A

Cardiac glycosides

These are steroidal saponin-like compounds possessing unsaturated lactone rings attached in the β position at C-17. Sugar residues are linked glycosidically via the C-3 OH groups of the steroid aglycones. The aglycones have a tetracyclic steroidal nucleus with OH groups at positions C-3 and C-14. Most plants that contain these compounds (and the compounds themselves) are restricted for use by medical practitioners only; however, all health practitioners should have a basic understanding of their pharmacology since they are still prescribed in general practice.

Aglycones are derived from mevalonic acid, but the final molecules arise from a condensation of a C_{21} steroid with a C_2 unit (**cardenolides**) or C_3 unit (**bufadienolides**).

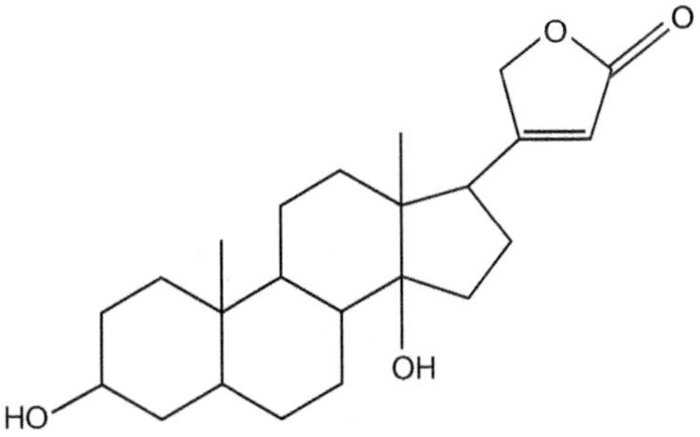

cardenolide structure

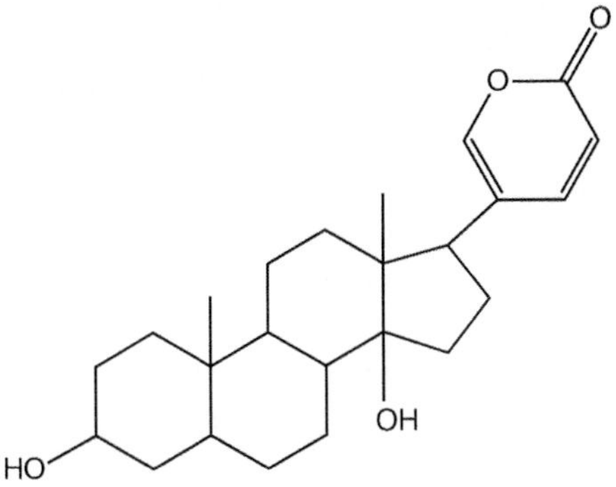

bufadienolide structure

Sugar moieties are composed of three sugar units: glucose, rhamnose and specific sugars such as digitoxose, which occur only in conjunction with cardiac glycoside. The sugar moiety confers on the glycoside solubility properties important in its absorption and ability to bind to heart muscle (El-Seedi *et al.*, 2019). The presence of OH groups increases the onset of action and subsequent dissipation from the body. Glycosides with few OH groups tend to be lipophilic and are absorbed and eliminated more slowly (Bruneton, 1995). The most widely used drug in this category is digoxin, which is actually a derivative of lantoside C, one of the glycosides in *Digitalis lanata* (Plantaginaceae) (Samuelsson, 1992).

A select group of plant families are known to produce cardiac glycosides, the most notable being the Asparagaceae, Plantaginaceae and Apocynaceae. Some of the more important of the glycosides along with their plant source are listed in Table 6.3.

Action of cardiac glycosides

Cardiac glycosides increase the force and speed of systolic contraction. In the failing heart they cause a more complete emptying of ventricles and shortening

Table 6.3. Plant sources of cardiac glycosides

Glycoside	Plant source	Family
Digitoxin, gitoxin	*Digitalis purpurea*	Plantaginaceae
Lanatosides A–E, digoxin	*Digitalis lanata*	Plantaginaceae
Convallotoxin, convalloside	*Convallaria majalis*	Asparagaceae
Stropanthin	*Strophanthus gratus*	Apocynaceae
Hellebrin	*Helleborus viridis*	Ranunculaceae
Proscillaridin, scillaren A	*Drimia maritima*	Asparagaceae
Odoroside, oleandrin	*Nerium oleander*	Apocynaceae

in length of systole. The heart has more time to rest between contractions. Increased cardiac output causes a lower heart rate and increases renal excretion. As to the pharmacology of digoxin itself, four main actions occur:

1. Positive inotropic effect – increases myocardial contractility due to direct inhibition of membrane-bound Na^+/K^+-ATPase leading to an increase in intracellular Ca^{2+} (i.e. Ca^{2+} replacing K^+) leading to an increased muscle contraction.
2. Increase in atrial and ventricular myocardial excitability – this may lead to arrhythmias.
3. Decrease in rate of atrioventricular conduction.
4. Increase in vagal tone and myocardial sensitivity to vagal impulses.

Toxicity

These compounds have a low therapeutic index (0.5), meaning the therapeutic dose is not much lower than the toxic dose. Digitalis intoxication affects the body in many ways. There are gastrointestinal symptoms (nausea, vomiting, diarrhoea), vision disturbances, neurological symptoms (headache, neuralgia, drowsiness) and cardiovascular symptoms including worsening cardiac failure and arrhythmias.

Disturbance of electrolytes occurs, and potassium may need to be administered. There are also many contraindications and drug interactions to be aware of. For a review of digoxin therapeutics and toxicology see El-Seedi *et al.* (2019).

References

Arao, T., Udayaina, M., Kinjo, J. and Nohara, T. (1998) Preventative effects of saponins from the *Peuraria lobata* root on *in vitro* immunological liver injury of rat primary hepatocyte cultures. *Planta Medica* 64(5), 413–416. doi: 10.1055/s-2006-957471

Asgarpanah, J. and Mohajerani, R. (2012) Phytochemistry and pharmacologic properties of *Urtica dioica* L. *Journal of Medicinal Plants Research* 6(46), 5714–5719. doi: 10.5897/JMPR12.540

Bruneton, J. (1995) *Pharmacognosy, Phytochemistry, Medicinal Plants.* Lavoisier Publications, Paris.

Budhiraja, R.D., Krishnan, K. and Sudhir, S. (2000) Biological activity of withanolides. *Journal of Scientific Research* 59, 904–911. Available at: http://nopr.niscair.res.in/bitstream/123456789/26628/1/JSIR%2059(11)%20904-911.pdf (accessed 26 August 2019).

Chen, L., He, H. and Qiu, F. (2010) Natural withanolides: an overview. *Natural Product Reports* 28, 705. doi: 10.1039/c0np00045k

Della Loggia, R., Tubaro, A., Sosa, S., Becker, H., Saar, St. and Isaac, O. (1994) The role of triterpenoids in the topical antiinflammatory activity of *Calendula officinalis* flowers. *Planta Medica* 60, 516–520.

Dewick, P.M. (2009) The mevalonate and methylerythritol phosphate pathways: terpenoids and steroids. In: Dewick, P.M. (ed.) *Medicinal Natural Products: a Biosynthetic Approach*, 3rd edn. Wiley, Chichester, UK, pp. 187–310. doi: 10.1002/9780470742761

El-Seedi, H., Khalifa, S.A.M., Taher, E.A., Farag, M.A., Saeed, A., *et al.* (2019) Cardenolides: insights from chemical structure and pharmacological utility. *Pharmacological Research* 141, 123–175. doi: 10.1016/j.phrs.2018

Favel, A., Steinmetz, M.D., Regli, P., Vidal-Ollivier, E., Elias, R. and Balansard, G. (1994) *In vitro* antifungal activity of triterpenoid saponins. *Planta Medica* 60(1), 50–53. doi: 10.1055/s-2006-959407

Fulder, S. (1980) *The Root of Being: Ginseng and the Pharmacology of Harmony.* Hutchinson, London.

Gao, J., Min, B., Ahn, E., Nakamura, N., Lee, H. and Hattori, M. (2002) New triterpene aldehydes, lucialdehydes A–C, from *Ganoderma lucidum* and their cytotoxicity against murine and human tumor cells. *Chemical and Pharmaceutical Bulletin* 50(6), 837–840.

Guclu-Ustundag, O. and Massa, G. (2007) Saponins: properties, applications and processing. *Critical Reviews in Food Science and Nutrition* 47, 231–258. doi: 10.1080/10408390600698197

Harborne, J. and Baxter, H. (1993) *Phytochemical Dictionary.* Taylor & Francis, London.

Hirano, T., Homma, M. and Oka, K. (1994) Effects of stinging nettle root extracts and their steroidal components on the Na$^+$, K$^{(+)}$-ATPase of the benign prostatic hyperplasia. *Planta Medica* 60(1), 30–33. doi: 10.1055/s-2006-959402

Hirotani, M., Tsutomu, F. and Shiro, M. (1985) Lanostane derivatives from *Ganoderma lucidum. Phytochemistry* 24, 2055–2061.

Hobbs, C. (1995) *Medicinal Mushrooms: an Exploration of Tradition, Healing and Culture.* Botanica Press, Santa Cruz, California.

Hu, C.-C., Lin, J.-T., Liu, S.-C. and Yang, D.-J. (2007) A spirostanol glycoside from wild yam (*Dioscorea villosa*) extract and its cytostatic activity on three cancer cells. *Journal of Food and Drug Analysis* 15(3), 310–315. Available at: http://lawdata.com.tw/File/PDF/J991/A04971503_310.pdf (accessed 3 December 2020).

James, J.T. and Dubery, I.A. (2009) Pentacyclic triterpenoids from the medicinal herb, *Centella asiatica* (L.) Urban. *Molecules* 14(10), 3922–3941. doi: 10.3390/molecules14103922

Komoda, Y., Sonada, Y. and Sate, Y. (1989) Gandoderic acid and its derivatives as cholesterol synthesis inhibitors. *Chemical Pharmaceutical Bulletin* 37(2), 531–533. Available at: https://www.jstage.jst.go.jp/article/cpb1958/37/2/37_2_531/_pdf (accessed 3 December 2020).

Kreis, W. and Müller-Uri, F. (2010) Biochemistry of sterols, cardiac glycosides, brassinosteroids, phytoecdysteroids and steroid saponins. In: Wink, M. (ed.) *Biochemistry of Plant Secondary Metabolism. Annual Plant Reviews* Volume 40. Blackwell Publishing, Oxford, UK, pp. 304–363. https://doi.org/10.1002/9781444320503.ch6

Lacaille-Dubois, M.A. and Wagner, H. (1996) A review of the biological and pharmacological activities of saponins. *Phytomedicine* 2, 363–386. doi: 10.1016/S0944-7113(96)80081-X

Lagarda, M.J., García-Llatas, G. and Farré, R. (2004) Analysis of phytosterols in foods. *Journal of Pharmaceutical and Biomedical Analysis* 41, 1486–1496.

Lee, D., Noh, H., Choi, J., Lee, K., Lee, M., *et al.* (2013) Anti-inflammatory cycloartane-type saponins of *Astragalus membranaceus*. *Molecules* 18, 3725–3732. doi: 10.3390/molecules18043725

Liu, X., Wang, J. and Yuan, J. (2005) Pharmacological and anti-tumor activities of *Ganoderma* spores processed by top-down approaches. *Journal of Nanoscience and Nanotechnology* 5(12), 1–13. doi: 10.1166/jnn.2005.448

Lou, J., Dimitrova, D.M., Murchison, C., Arnold, G.C., Belding, H., *et al.* (2018) *Centella asiatica* triterpenes for diabetic neuropathy: a randomized, double-blind, placebo-controlled, pilot clinical study. *Esper Dermatology* 20(2) (Suppl. 1), 12–22. doi: 10.23736/S1128-9155.18.00455-7

Morigiwa, A., Kitabatake, K., Fujimoto, Y. and Ikehaura, N. (1986) Angiotensin converting enzyme-inhibitory triterpenes from *Ganoderma lucidum*. *Chemical Pharmaceutical Bulletin* 34(7), 3025–3028. doi: 10.1248/cpb.34.3025

Moses, T., Kalliope, K. and Papadopoulou, A. (2014) Metabolic and functional diversity of saponins, biosynthetic intermediates and semi-synthetic derivatives. *Critical Reviews in Biochemical and Molecular Biology* 49(6), 439–462. doi: 10.3109/10409238.2014.953628

Muley, B.P., Khadabadi, S.S. and Banarase, N.B. (2009) Phytochemical constituents and pharmacological activities of *Calendula officinalis* Linn (Asteraceae): a review. *Tropical Journal of Pharmaceutical Research* 8(5), 455–465. doi: 10.4314/tjpr.v8i5.48090

Oh, S.R., Kinjo, J., Shii, Y., Ikeda, T., Nohara, T., *et al.* (2000) Effects of triterpenoids from *Pueroria lobata* on immunohemolysis. *Planta Medica* 66(6), 506–510. doi: 10.1055/s-2000-8614

Osbourn, A.E. (2003) Saponins in cereals. *Phytochemistry* 62(1), 1–4. https://doi.org/10.1016/S0031-9422(02)00393-X

Panossian, A. (2017) Understanding adaptogenic activity: specificity of the pharmacological action of adaptogens and other phytochemicals. *Annals of the New York Academy of Sciences* 1401(1), 49–64. https://doi.org/10.1111/nyas.13399

Patel, K., Gadewar, M., Tahilyani, V. and Patel, D. (2012) A review on pharmacological and analytical aspects of diosgenin: a concise report. *Natural Product Bioprospecting* 2(2), 46–52.

Recio, M.C., Giner, R., Manez, M. and Rios, J.L. (1995) Structural requirements for the anti-inflammatory activity of natural terpenoids. *Planta Medica* 61(2), 182–185. doi: 10.1055/s-2006-958045

Ren, S., Zhang, H., Mu, Y., Sun, M. and Liu, P. (2013) Pharmacological effects of Astragaloside IV: a literature review. *Journal of Traditional Chinese Medicine* 33(3), 413–416. doi: 10.1016/s0254-6272(13)60189-2

Ren, Y. and Kinghorn, A.D. (2019) Natural product triterpenoids and their semi-synthetic derivatives with potential anticancer activity. *Planta Medica* 85(11/12), 802–814. doi: 10.1055/a-0832-2383

Samuelsson, G. (1992) *Drugs of Natural Origin*. Swedish Pharmaceutical Press, Stockholm.

Sawai, S. and Saito, K. (2011) Triterpenoid biosynthesis and engineering in plants. *Frontiers in Plant Science* 2, 25. https://doi.org/10.3389/fpls.2011.00025

Shah, R., Gulati, V. and Palombo, E.A. (2012) Pharmacological properties of guggulsterones, the major active components of gum guggul. *Phytotherapy Research* 26(11), 1594–1605. https://doi.org/10.1002/ptr.4647

Son, I.S., Kim, J.H., Sohn, H.Y., Son, K.H., Kim, J.-S. and Kwon, C.-S. (2007) Antioxidative and hypolipidemic effects of diosgenin, a steroidal saponin of yam (*Dioscorea* spp.) on high-cholesterol fed rats. *Bioscience, Biotechnology, and Biochemistry* 71(12), 3063–3071. https://doi.org/10.1271/bbb.70472

Szakiel, A., Ruszkowski, D. and Janiszowska, W. (2005) Saponins in *Calendula officinalis* L. – structure, biosynthesis, transport and biological activity. *Phytochemical Reviews* 4, 151–158.

Takechi, M. and Tanaka, Y. (1995) Haemolytic time course differences between steroid and triterpenoid saponins. *Planta Medica* 61(1), 76–77. doi: 10.1055/s-2006-958006

Urbanek, T. (2017) The clinical efficacy of *Ruscus aesculatus* extract: is there enough evidence to update the pharmacotherapy guidelines for chronic venous disease? *Phlebological Review* 25(1), 75–80. https://doi.org/10.5114/pr.2017.70594

Vincken, J. Heng, L. de Groot, A. and Gruppen, H. (2007) *Saponins, classification and occurrence in the plant kingdom. Phytochemistry* 68, 275–297.

Wahlstrom, M. (1987) *Adaptogens: Nature's Key to Well-being*. Utgivare, Goteborg, Sweden.

Woyengo, T.A., Ramprasath, V.R. and Jones, P.J.H. (2009) Anticancer effects of phytosterols. *European Journal of Clinical Nutrition* 63, 813–820. https://doi.org/10.1038/ejcn.2009.29

Yen, M.-H., Lin, C.-C., Chuang, C.-H. and Lin, S.-C. (1994) Anti-inflammatory and hepatoprotective activity of saikosaponin-f and the root extract of *Bupleurum kaoi*. *Fitoterapia LXV*, 409–412.

Zhu, Q., Bang, T.H., Ohnuki, K., Sawai, T., Sawai, K. and Shimizu, K. (2015) Inhibition of neuraminidase by *Ganoderma* triterpenoids and implications for neuraminidase inhibitor design. *Scientific Reports* 5, 13194. doi: 10.1038/srep13194

Resins and Cannabinoids

Introduction

Resins are solid, brittle substances secreted by plants into special ducts, often as a response to damage to the plant by wounding, wind or insect damage. Their main role appears to be protection of the plant from attack by fungi and insects. Resins are difficult to classify because of their amorphous nature; they are complex mixtures that include a diverse set of constituents including any combination of lignans, resin acids, resin alcohols, resinotanninols, cannabinoids, esters and resenes (see Table 7.1).

Resins are insoluble in water but soluble in alcohol and fixed oils. They are heavier than water and volatile oils, with high boiling points. Resins are translucent and burn with a characteristic smoky flame – hence their use in incense. They have fixative actions, making them useful ingredients in crafts and industry.

While classifying individual resins is difficult, they are sometimes classified as mixtures with other plant constituents, for example gum-resins, oleo-gum-resins, glycoresins. One of the most well-known resins called **rosin** comes from the *Pinus* genus (Pinaceae). This amber-coloured resin is mainly used in varnishes and other industrial products.

Major resin- and oleo-gum-resin-containing herbs

Myrrh

Oleo-gum-resin exudes from incisions made in bark of myrrh trees in North Africa (*Commiphora molmol* and *Commiphora myrrha*, Burseraceae) and India (*Commiphora mukul*), where it is known as guggul. The main constituents are: resin (25–40%), gum (60%) and volatile oil (2.5–8%) along with a bitter principle. Myrrh's characteristic odour is derived from furanosesquiterpenes

 DOI: 10.1079/9781789243079.0007

Table 7.1. Classification of resins

Class	Composition	Examples and plant source
Pure resins	Solid complexes	Guaiac (*Guaiacum officinale*)
		Hashish (*Cannabis sativa*)
Oleoresins	Mixture of resins and volatile oils	Capsaicin (*Capsicum frutescens*)
		Gingerols (*Zingiber officinale*)
		Mastic (*Pistacia lentiscus*)
		Turpentine (*Pinus palustris*)
		Aspidinol (*Dryopteris filix-mas*)
		Balm of Gilead (*Populus balsamifera*)
Oleo-gum-resins	Mixture of resins, volatile oils and gums	Frankincense (*Boswellia carterii*)
		Asofoetida (*Ferula assa-foetida*)
Balsams	Cinnamic or benzoic acids and their esters	Benzoin (*Styrax benzoin*)
		Balsam of Tolu (*Myroxylon balsamum*)
		Storax (*Liquidambar orientalis, Liquidambar styraciflua*)
Resin acids	Diterpenoid acids	Myrrh (*Commiphora molmol*)
Resinotanninols	Complex alcohols	Asaresinatannols (*F. assa-foetida*)
Resin alcohols	Cannabinoids	Tetrahydrocannabinol (THC) and cannabidiol (CBD) (*C. sativa*)
Glycoresins	Sugars, resin acids	Podophyllin (*Podophyllum peltatum*)
		Jalap (*Exogonium purga*)

including **furanoeudesma-1,3-diene** (Hanus *et al.*, 2005). Myrrh has a long tradition of use for perfumes and incense.

Its actions are antiseptic, antimicrobial, astringent and stimulant and it is widely used for treatment of gingivitis, throat infections and wounds.

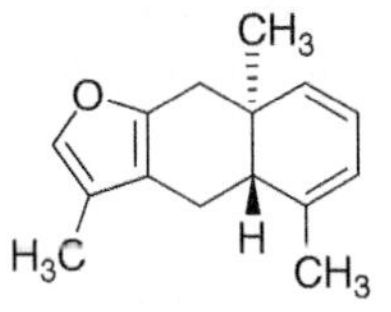

furanoeudesma-1,3-diene

Ginger

The oleoresin from the rhizome of ginger, *Zingiber officinale* (Zingiberaceae), consists of 20–25% essential oil and 25–30% pungency principles (Zachariah, 2008). These are phenolic arylalkanones known as **gingerols**, derived from phenylpropanes, but with extended hydrocarbon chains. Gingerols are reduced to the more pungent **shogaols** during drying, while cooking converts gingerols to the less pungent but more flavoursome ketone, **zingerone** (van Wyk, 2013). **Gingerenones** are diarylheptenones formed from dimerization of gingerols.

The most common component of ginger essential oil is **zingiberene**, a mono-cyclic sesquiterpenoid, also the key flavour ingredient in ginger. Other flavour components in ginger include the sesquiterpenoid, α-curcumene also known as ar-curcumene, and monoterpenoids such as: camphene, β-phellandrene, 1,8-cineole and citral (Koroch *et al.*, 2007). Australian ginger oil has a high (up to 19%) content of citral, giving the spice a more citrus-like flavour (Zachariah, 2008).

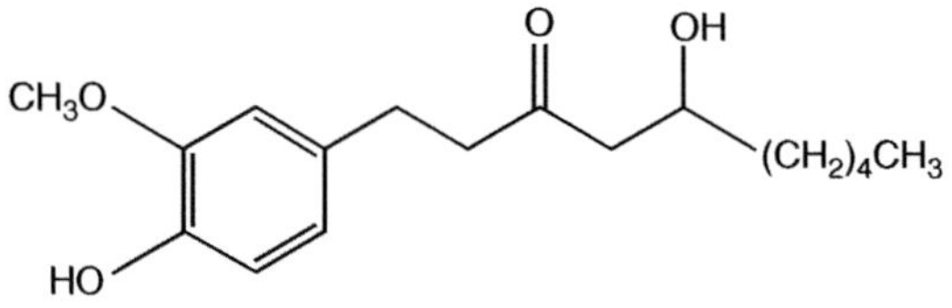

6-gingerol – from ginger oleoresin

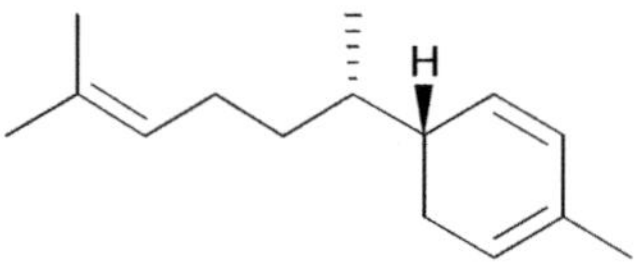

zingiberene

6-Gingerols are anti-inflammatory, in part due to inhibitory effect on iNOS (inducible nitric oxide synthase) expression (Tsai *et al.*, 2005). Other attributed actions include antibacterial, antitumour, antiplatelet, antinauseant, antiulcer and analgesic (Zachariah, 2008).

The diarylheptanoid **gingerenone A** is another bioactive con-stituent. *In vivo* studies reveal that this compound has anti-obesity effects due to suppression of adipocytes and inhibition of macrophage infiltration (Suk *et al.*, 2017). In addition, this compound prevents atherosclerosis formation by in-hibition of proteins via the NF-κB (nuclear factor kappa B) pathway (Kim *et al.*, 2018). Ginger contains numerous antibacterial compounds, and of eight com-pounds tested, gingerenone A and shogaol were the most potent inhibitors of SaHPPK, an antimicrobial target involved in folate synthesis of *Staphylococcus aureus* (Rampogu *et al.*, 2018).

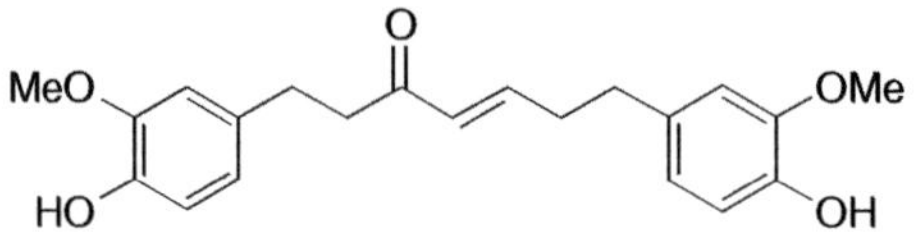

gingerenone A

Asafoetida

Asafoetida (hing), one of the most pungent of all spices, is obtained from the rhizomes and roots of a shrub *Ferula assa-foetida* (Apiaceae) native to the south-west Asian region. Rich in an oleo-gum-resin, asafoetida is made up of 6–17% volatile oils, 40–64% resin and 7–25% gum. The volatile oil contains disulfide compounds like those found in garlic, along with monoterpenes such as α- and β-pinene. The resin consists of **farnesiferols** A and B (sesquiterpe-noid coumarins), ferulic acid, umbelliferone and asaresinatannols (Mahendra and Bisht, 2012). Properties of hing include antifungal, antispasmodic, expec-torant, stimulant, emmenagogue and vermifuge, while it reputedly acts as an antidote to opioids (Mahendra and Bisht, 2012).

Capsicum

Capsaicin is the pungent principle derived from fruits of cayenne and chilli peppers (*Capsicum annum*; *C. frutescens*, Solanaceae). Capsaicin, a phenolic amide, is present in the fruit at a level of only 0.02%, yet its taste is detectable even in minute doses. The compound acts as a local anaesthetic and pain re-liever through a complex mechanism (see Chapter 11, this volume).

Podophyllum

Podophyllin is a resinous mixture derived from dried rhizomes and roots of *Podophyllum hexandrum* and *P. peltatum* (Berberidaceae). The resin is domin-ated by lignans, most notably **podophyllotoxin** and **a and b peltatin**, with small quantities of lignan glycosides. *P. hexandrum*, native to the Himalayan region of Asia, is the traditional source of podophyllotoxin – a widely used pharmaceutical drug – and overharvesting of this species has led to it being classified as endangered. The problem is being countered by cultivation of both Asian and American species for production of the resin, using *in vitro* methods, careful selection of chemotypes, and innovative extraction methods (Moraes *et al.*, 2002; Bedir *et al.*, 2006; Sharma *et al.*, 2017).

Podophyllin powder has a peculiar bitter taste and is highly irritating to mucous membranes, especially of the eyes. Its caustic nature is utilized in the form of topical applications for warts including condylomas (genital warts) (Li *et al.*, 2012). Internally it acts as a drastic though slow-acting purgative. Podophyllin is used in the semi-synthesis of the anticancer drugs etoposide and teniposide. These drugs are antimitotic – they stop cell divisions – and are used for treatment of lung, testicular and other types of cancer (Chaudhari *et al.*, 2014; Sharma *et al.*, 2017).

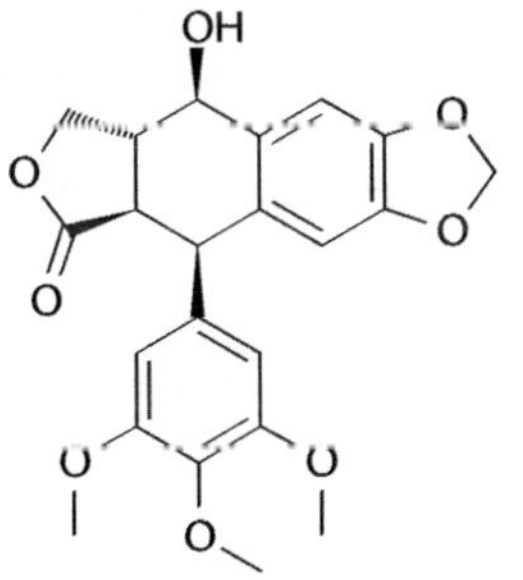

podophyllin

Mastic

This oleo-gum-resin from *Pistachia lentiscus* (Anacardiaceae) has been shown to have antimicrobial and antioxidant properties, and is often used as an ingredient in cosmetics (Assimopoulou *et al.*, 2005).

Balsams

These are resinous mixtures that contain cinnamic and/or benzoic acid or their esters. **Benzoin** is a balsamic resin derived from deep incisions made into the bark of *Styrax benzoin* (Styracaceae) and related species of trees in South-east Asia. It has been used like incense in ritual ceremonies, as a perfume and fixative. Benzoin contains cinnamic, benzoic and oleanane triterpenic acids. Its action is antiseptic, stimulant, expectorant, diuretic and antifungal. Benzoin is used as a food preservative and as an ingredient in pharmaceutical preparations such as Whitfield's ointment (with salicylic acid), for ringworm and athlete's foot (Tyler *et al.*, 1988; Hovaneissian *et al.*, 2008).

   Storax, with similar uses to benzoin, derives from *Liquidambar orientalis* (Altingiaceae) and other species. Storax is a chemical mixture, containing cinnamyl alcohol, ethyl cinnamate, cinnamyl acetate and a complex mixture of essential oil components (Kartal *et al.*, 2012). Traditionally the balsam has been used to treat inflammation, stomach pain, bronchitis, liver disorders and externally for skin disorders such as eczema, scabies and leukoderma, while research has revealed sedative and anticonvulsive activity (El-Readi *et al.*, 2013). Antifungal properties help make storax an effective useful wood preservative, while the constituent cinnamyl alcohol was shown to repel termites (Kartal *et al.*, 2012).

Cannabinoids

Cannabis resin, or Indian hemp, is derived from the dried flowering tops of pistillate plants of *Cannabis sativa* (Cannabaceae), now regarded as the only *Cannabis* species (McPartland, 2018). The plants contain 15–20% resin consisting mainly of Δ^9-**tetrahydrocannabinol**, or THC, and **cannabidiol** (CBD) along with related cannabinoids. This class of C_{21} compounds, found almost exclusively in *C. sativa*, are benzotetrahydropyrans, their structures derive from both acetyl-CoA and MEP (methylerythritol phosphate – monoterpenoid) pathways. They are synthesized within glandular trichomes (hairs) of flowers and stored at the tips of the trichomes.

Ratios of THC and CBD vary according to whether the plant has been bred for recreational use, with higher THC content, (so-called 'drug-type') or wild types such as those used for hemp production ('fibre-type'), which are higher in CBD. The ratio is dictated by the expression of enzymes that modify the initial cannabinoid metabolite, **cannabigerolic acid** (CBGA), to form either THC or CBD acid. The over 100 known cannabinoids have been categorized into several structural classes (Table 7.2). These classes contain one or more characteristic acids, which readily undergo decarboxolation (a non-enzymic process) to form the key active molecules, such as THC and CBD (Flores-Sanchez and Verpoorte, 2008; Appendino *et al.*, 2011).

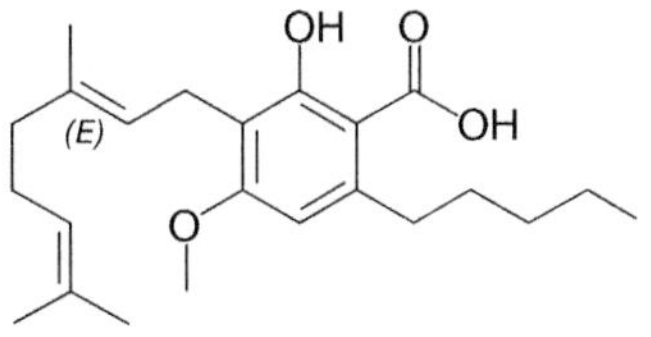

cannabigerolic acid[a]

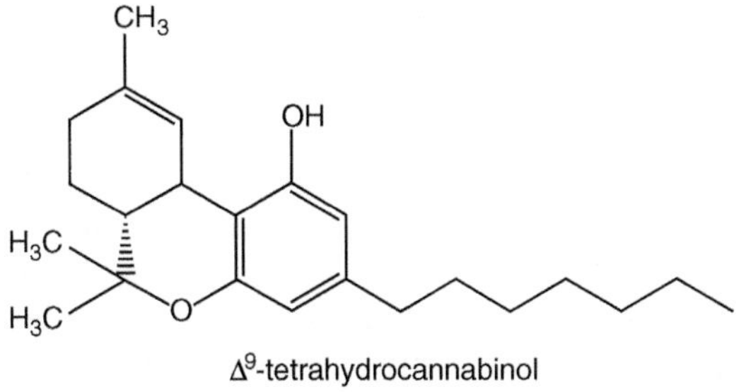

Δ^9-tetrahydrocannabinol

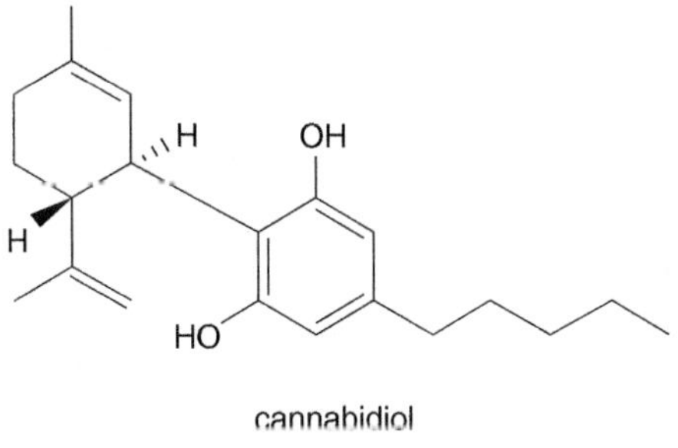

cannabidiol

Table 7.2. Major cannabinoid structural classes and their actions (from Brenniesen, 2007; Appendino *et al.*, 2011)

Cannabinoid class	Representative compounds	Therapeutic actions
Cannabigerol	Cannabigerolic acid (CBGA)	Antibacterial, anti-inflammatory, muscle relaxant
Cannabichromene	Cannabichromenic acid (CBCA) Cannabichromene (CBC) Cannabicylol	Antibacterial, anti-inflammatory, analgesic
Cannabidiol	Cannabidiol (CBD) Cannabinmovone	Anxiolytic, analgesic, anti-inflammatory, spasmolytic, neuroprotective, antioxidant, antitumour, antidepressant
Δ^9-Tetrahydrocannibol	Δ^9-Tetrahydrocannibol (THC) Tetrahydrocannibolic acid (THCA) Δ^8-Tetrahydrocannibol Cannabinol (CBN)	Euphoriant, analgesic, antiemetic, neuroprotective, antibacterial, anti-inflammatory, bronchodilator

Endocannabinoids

The discovery of cannabinoid receptors in the human brain in 1988 led scientists to search for endogenous molecules that bind to these receptors, in a similar way that endorphins had been found to bind to opioid receptors. The first such compound discovered was the lipid **arachidonoyl-ethanolamide** or anandamide, an amide of arachidonic acid that binds to CB_1 receptors, and which produces similar biological effects as phytocannabinoids. Anandamide is derived from polyunsaturated fatty acids, and structurally related to the alkylamide group (see Chapter 11, this volume).

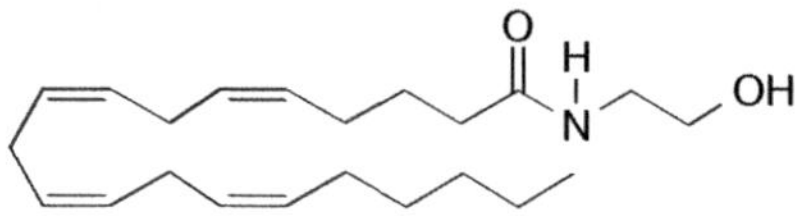

anandamide – an endocannabinoid

Following the discovery of a second cannabinoid receptor (CB_2) located mostly outside of the central nervous system, another endocannabinoid, **2-arachidonoyl glycerol (2-AG)** was isolated with affinity for both CB_1 and CB_2 receptors.

Endocannabinoids were defined in the journal *Nature Neuroscience* as endogenous compounds that can directly activate or block cannabinoid CB_1 and/or

CB_2 (Mechoulam *et al.*, 2014), and are representative of a family of lipid mediators (signalling molecules) found both within the brain as well as in peripheral tissues (Pasquariello *et al.*, 2009). They monitor most aspects of brain function including pain control, learning and memory, appetite and reward (Kennedy, 2014).

Cannabinoid therapeutics

Despite some differences, there are numerous pharmacological and therapeutic actions shared by the cannabinoids in *C. sativa*, and these actions correlate closely with the known actions of endocannabinoids. However, not all cannabinoid actions are mediated via CB_1 and CB_2. Notably CBD binds only weakly to CB receptors, and it may in fact counteract some of the psychoactive effects of THC. There has been a strong trend towards the medical use of CBD and CBD chemotypes of *C. sativa*, due to their lack of psychoactivity, which may improve patient compliance. However, given the potential for synergistic and additive actions between cannabinoids, and the moderating effect on CBD on the psychoactive effects of THC (Kennedy, 2014), there would appear to be therapeutic benefits in using whole plant extracts with low CBD:THC ratios (Russo and Guy, 2006). Whole plant extracts also have the advantage of containing the volatile mono- and sesquiterpenes (see Chapter 8, this volume), which further extends the therapeutic range for *C. sativa*, due to the purported 'entourage effect' (Lewis *et al.*, 2018).

Cannabichromene (CBC) occurs mainly in drug-type *C. sativa*, where it has been found to potentiate the anti-inflammatory effects of THC through an additive action, while producing anti-inflammatory and analgesic effects in its own right (DeLong *et al.*, 2010). For THC itself, significant analgesic action comes from inhibition of the neurotransmitter GABA (γ-aminobutyric acid) in a similar manner to opioids, though unlike opioids they do not inhibit postsynaptic neurones. Anandamide produces analgesia through the same mechanism, however, it is less potent than THC (Vaughan and Christie, 2002). CBD has demonstrated a broad range of central nervous system effects in addition to analgesia, including: immunosuppressant, anti-inflammatory, anti-allergic, improvement of mood, appetite stimulation, lowering of intraocular pressure in glaucoma, neuroprotection and antineoplastic. Some of these effects come from binding to serotonin (5-HT 1A), opioid and dopamine receptors (Scuderi *et al.*, 2009). In a recent study, inhaled cannabis was reported to give significant relief for 50% of nearly 2000 headache and migraine sufferers, however, there was little correlation with THC:CBD ratios of the strains that were most effective (Cutler *et al.*, 2019). Other notable conditions for which CBD and cannabinoids, in general, are being used are Parkinsonism and epilepsy.

All five major cannabinoids (CBD, CBC, cannabigerol (CBG), 9-THC and cannabinol (CBN)) showed potent antibacterial activity against methicillin-resistant *Staphylococcus aureus* (MRSA), with minimum inhibitory concentration (MIC) values in the range of 0.5–2 µg/ml (Appendino *et al.*, 2011). THC, CBD and CBG have demonstrated a non-cannabinoid-mediated inhibition of the proliferation of a hyper-proliferating human keratinocyte cell line, pointing to therapeutic potential for treating psoriasis (Wilkinson and Williamson, 2007).

Safety issues

Most of the safety concerns for *C. sativa* are associated with the psychotropic component THC. Being a lipophilic compound, THC is quickly absorbed, but slowly excreted. Symptoms include euphoria, fast pulse, loss of coordination and mild hallucinations, while in high doses, anxiety and bradycardia may occur. Chronic use may involve personality changes, short-term memory loss and, in susceptible individuals, psychosis and schizophrenia (Wink and van Wyk, 2008). Dependency resulting from regular cannabis use is estimated at approximately 9% of users, a figure lower than for key addictive drugs such as tobacco and heroin (Kennedy, 2014).

CBD lacks the psychotropic effects of THC. It is regarded as having a more favourable safety profile (Iffland and Grotenhermen, 2017).

References

Appendino, G., Chianese, G. and Taglialatela-Scafati, O. (2011) Cannabinoids: occurrence and medicinal chemistry. *Current Medicinal Chemistry* 18(7), 1085–1099. doi: 10.2174/092986711794940888

Assimopoulou, A.N., Zlatanos, S.N. and Papageorgiou, V.P. (2005) Antioxidant activity of natural resins and bioactive triterpenes in oil substrates. *Food Chemistry* 92, 721–727. https://doi.org/10.1016/j.foodchem.2004.08.033

Bedir, E., Tellez, M., Lata, H., Khan, I., Cushman, K.E. and Moraes, R.M. (2006) Post-harvest and scale-up extraction of American mayapple leaves for podophyllotoxin production. *Industrial Crops and Products* 24(1), 3–7. https://doi.org/10.1016/j.indcrop.2005.10.001

Brenniesen, R. (2007) Chemistry and analysis of phytocannabinoids and other cannabis constituents. In: Elsohly, M.A. (ed.) *Forensic Science and Medicine: Marihuana and the Cannabinoids*. Humana Press, Totowa, New Jersey, pp. 17–49. https://doi.org/10.1007/978-1-59259-947-9_2

Chaudhari, S.K., Bibi, Y. and Arshad, M. (2014) *Podophyllum hexandrum*: an endangered medicinal plant from Pakistan. *Pure Applied Biology* 3(1), 19–24. doi: 10.19045/bspab.2014.31003

Cutler, C., Spradlin, A., Cleveland, M.J. and Craft, R.M. (2019) Short- and long-term effects of cannabis on headache and migraine. *Journal of Pain* 21(5–6), 722–730. https://doi.org/10.1016/j.jpain.2019.11.001

DeLong, G.T., Wolf, C.E., Poklis, A. and Lichtman, A.H. (2010) Pharmacological evaluation of the natural constituent of *Cannabis sativa*, cannabichromene and its modulation by Δ⁹-tetrahydrocannabinol. *Drug and Alcohol Dependence* 112(1–2), 126–133. doi: 10.1016/j.drugalcdep.2010.05.019

El-Readi, M.Z., Eid, H.H., Ashour, M.L., Eid, S.Y., Labib, R.M., *et al.* (2013) Variations of the chemical composition and bioactivity of essential oils from leaves and stems of *Liquidambar styraciflua* (Altingiaceae). *Journal of Pharmacy and Pharmacology* 65, 1653–1663. doi: 10.1111/jphp.12142

Flores-Sanchez, I.J. and Verpoorte, R. (2008) Secondary metabolism in cannabis. *Phytochemistry Reviews* 7, 615–639. doi: 10.1007/s11101-008-9094-4

Hanus, L.O., Řezanka, T., Dembitsky, V.M. and Arieh Moussaieff, A. (2005) Myrrh – *Commiphora* chemistry. *Biomedical Papers* 149(1), 3–28. doi: 10.5507/bp.2005.001

Hovaneissian, M., Archier, P., Mathe, C., Culio, G. and Vieillescazes, C. (2008) Analytical investigation of styrax and benzoin balsams by HPLC-PAD-fluorimetry and GC-MS. *Phytochemical Analysis* 19, 301–310. doi: 10.1002/pca.1048

Iffland, K. and Grotenhermen, F. (2017) An update on safety and side effects of cannabidiol: a review of clinical data and relevant animal studies. *Cannabis and Cannabinoid Research* 2(1), 139–154. doi: 10.1089/can.2016.0034

Kartal, S.N., Terzi, E., Yoshimura, T., Arango, R., Clausen, C.A. and Green III, F. (2012) Preliminary evaluation of storax and its constituents: fungal decay, mold and termite resistance. *International Biodeterioration and Biodegradation* 70, 47–54. https://doi.org/10.1016/j.ibiod.2012.02.002

Kennedy, D.O. (2014) *Plants and the Human Brain.* Oxford University Press, Oxford, UK.

Kim, H.J., Son, J.E., Kim, J.H., Lee, C.C., Yang, H., *et al.* (2018) Gingerenone A attenuates monocyte-endothelial adhesion via suppression of I kappa B kinase phosphorylation. *Journal of Cellular Biochemistry* 119(1), 260–268. doi: 10.1002/jcb.26138

Koroch, A., Ranarivelo, L., Behra, O., Juliani, H.R. and Simon, J.E. (2007) Quality attributes of ginger and cinnamon essential oils from Madagascar. In: Janick, J. and Whipkey, A. (eds) *Issues in New Crops and New Uses.* American Society of Horticultural Science (ASHS) Press, Alexandria, Virginia, pp. 338–341. Available at: https://hort.purdue.edu/newcrop/ncnu07/pdfs/koroch338-341.pdf (accessed 4 December 2020).

Lewis, M.A., Russo, E.B. and Smith, K.M. (2018) Pharmacological foundations of Cannabis chemovars. *Planta Medica* 84(4), 225–233. doi: 10.1055/s-0043-122240

Li, M., Zhou, L., Yang, D., Li, T. and Li, W. (2012) Biochemical composition and antioxidant capacity of extracts from *Podophyllym hexandrum* rhizome. *BMC Complementary and Alternative Medicine* 12, 263. doi: 10.1186/1472-6882-12-263

Mahendra, P. and Bisht, S. (2012) *Ferula asafoetida*: traditional uses and pharmacological activity. *Pharmacognosy Reviews* 6(12), 141–146. doi: 10.4103/0973-7847.99948

McPartland, J.M. (2018) Cannabis systematics at the levels of family, genus, and species. *Cannabis and Cannabinoid Research* 3(1), 203–212. doi: 10.1089/can.2018.0039

Mechoulam, R., Hanuš, L.O., Pertwee, R. and Howlett, A.C. (2014) Early phytocannabinoid chemistry to endocannabinoids and beyond. *Nature Neuroscience* 15(11), 757–764. doi: 10.1038/nrn3811

Moraes, R.M., Dayan, F.E. and Canel, C. (2002) The lignans of *Podophyllum*. In: Atta-ur-Rahman (ed.) *Studies in Natural Products Chemistry*, Volume 26. Bioactive Natural Products (Part G). *Studies in Natural Products Chemistry*. Elsevier Science, Amsterdam, pp. 149–182.

Pasquariello, N., Oddi, S., Malaponti, M. and Maccarrone, M. (2009) Regulation of gene transcription and keratinocyte differentiation by anandamide. In: Litwak, G. (ed.) *Vitamins and Hormones*, Volume 81. Academic Press, Cambridge, Massachusetts, pp. 441–467.

Rampogu, S., Baek, A., Gajula, R.G., Zeb, A., Bavi, R.S., *et al.* (2018) Ginger (*Zingiber officinale*) phytochemicals-gingerenone-A and shogaol inhibit SaHPPK: molecular docking, molecular dynamics simulations and *in vitro* approaches. *Annals of Clinical Microbiology and Antimicrobials* 17(1), 16. doi: 10.1186/s12941-018-0266-9

Russo, E. and Guy, G.W. (2006) A tale of two cannabinoids: the therapeutic rationale for combining tetrahydrocannabinol and cannabidiol. *Medical Hypotheses* 66, 234–246. doi: 10.1016/j.mehy.2005.08.026

Scuderi, C., De Filippis, D., Iuvone, T., Blasio, A., Steardo, A. and Esposito, G. (2009) Cannabidiol in medicine: a review of its therapeutic potential in CNS disorders. *Phytotherapeutic Research* 23, 597–602. doi: 10.1002/ptr2625

Sharma, P., Verma, A.K., Chauhan, S., Kharwar, S. and Shrivastava, P. (2017) Review on podophyllotoxin: sources, extraction, applications and current perspectives. *World Journal of Pharmaceutical Research* 6(1), 370–385.

Suk, S., Kwon, G.T., Lee, E., Jang, W.J., Yang, H., *et al.* (2017) Gingerenone A, a polyphenol present in ginger, suppresses obesity and adipose tissue inflammation in high-fat diet-fed mice. *Molecular Nutrition & Food Research* 61(10). doi: 10.1002/mnfr.201700139

Tsai, T.-H., Tsai, P.-J. and Ho, S.-C. (2005) Antioxidant and anti-inflammatory activities of several commonly used spices. *Journal of Food Science* 70(1). https://doi.org/10.1111/j.1365-2621.2005.tb09028.x

Tyler, V., Brady, J. and Robbers, J. (1988) *Pharmacognosy*, 9th edn. Lea & Febiger, Philadelphia, Pennsylvania.

Van Wyk, B. (2013) *Culinary Herbs and Spices of the World*. University of Chicago Press, Chicago, Illinois.

Vaughan, C.W. and Christie, M.J. (2002) Mechanisms of cannabinoid analgesia. In: Grotenhermen, F. and Russo, E.B. (eds) *Cannabis and Cannabinoids*. Haworth Integrative Healing Press, New York.

Wilkinson, J.D. and Williamson, E.M. (2007) Cannabinoids inhibit human keratinocyte proliferation through a non-CB_1/CB_2 mechanism and have a potential therapeutic value in the treatment of psoriasis. *Journal of Dermatological Science* 45(2), 87–92. doi: 10.1016/j.jdermsci

Wink, M. and van Wyk, B. (2008) *Mind-altering and Poisonous Plants of the World*. Briza Publications, Pretoria, South Africa.

Zachariah, T.J. (2008) Ginger. In: Parthasarathy, V.A., Chempakam, B. and Zachariah, T.J. (eds) *Chemistry of Spices*. CAB International, Wallingford, UK, pp. 70–96.

Essential Oils

Introduction

Essential oils are odorous exudations or principles stored in special plant cells – glands, glandular hairs, oil ducts or resin ducts – situated in any part of a plant. These oils are responsible for the distinctive aromas associated with individual plant species. They are soluble in alcohol and fats, but only slightly soluble in water. Most essential oils are colourless, a notable exception being azulene, which is blue. On exposure to light and air they readily oxidize. Essential oils are also called volatile oils, since they evaporate when subjected to heat.

Extraction of oils

Steam distillation is the predominant extraction method used. These and other methods of extraction are described and illustrated in *The Essential Oil Maker's Handbook* (Malle and Schmickl, 2012). Distillation is a method that involves the evaporation and subsequent condensation of liquids in order to produce, refine and concentrate essential oils. High quality oils are distilled once only, while some commercial oils are 'purified' by double or even triple distillation methods. Steam distillation is only effective for compounds whose molecular weights are below 300 Daltons (Sell, 2010).

Other extraction methods include:

- Solvent extraction – plants are heated in an organic solvent such as hexane and petroleum ether, after which the solvent is evaporated off. However, traces of solvent residues remain in the extracted oil, and the process is not suitable for medicinal grade oils, or for use in aromatherapy.
- Supercritical carbon dioxide extraction – a modern technique, in which plant materials are submitted to carbon dioxide under very high pressure. The resultant oils are free from solvent residues, and generally attain a high quality comparable to steam distillation. However, the high cost of production has limited their widespread use (Guba, 2018).

DOI: 10.1079/9781789243079.0008

- Enfleurage – a method restricted to a few delicate flowers, such as rose, jasmine and orange blossom (neroli). The process involves placing the flowers in a fat – shea butter, tallow and coconut oil can be used – between two glass plates separated by a wooden frame.
- Infusion – as for enfleurage, oils and fats make good solvents for essential oils. Aromatic plants can be macerated in vegetable oils, and once they are strained off the resultant oil is useful for topical applications. Chamomile, *Calendula* and *Hypericum* infused oils can be made by this method.
- Concrètes – a concrète is a semi-solid mass of plant material (flowers or other plant parts) extracted using a hydrocarbon solvent (e.g. petroleum ether or benzene). The solvent is recovered later by vacuum distillation.
- Resinoids – extracts from plant exudates, prepared as in concrètes.
- Absolutes – extracts from concrètes, resinoids or fat extracts using ethyl alcohol; waxes and fats are eliminated in the process, and the ethyl alcohol is later removed by vacuum distillation.

The olfactory system and our sense of smell

The olfactory nerves connect directly to the limbic system in the brain, and thereby influence sensory functions such as hunger, sex and emotions. Smelling involves the inhalation of microscopic chemicals such as those contained in essential oils, which flow through our nostrils into the nasal cavity. These molecules pass over moist bony structures called turbinates, to reach chemosensory olfactory neurones where odorant receptors are expressed. Here the chemical dissolves and comes in contact with a fine layer of hairs, which then stimulate the olfactory bulb of the brain (Wrigley and Fagg, 1990; Zozulza *et al.*, 2001).

The average human can discriminate between 4000 and 10,000 odour molecules, over 300 from the smell of a rose. Odour molecules dissolve in mucus that covers cilia in our nostrils. Dissolved molecules trigger olfactory receptors in epithelial sensory neurones – sending signals to the limbic system in the region of the brain stem (Assar, 2007).

Research into the chemistry of plant odours shows there is a definite link between the shape of molecules and their smell. Compounds that occur as enantiomers produce odours that differ according to which isomer is present, so that compounds with the same chemical formula can have different odours and different biological actions (Pavia *et al.*, 1982).

Chemistry of essential oils

Essential oils may be stored in secretory hairs or trichomes found on the surface of leaves or other plant parts, alternatively they are found in epidermal hairs (lysigenous oil sacs, excretion hairs or glands) or secretory cells within the plant. Other repositories of essential oils include nectaries and osmophores (floral fragrance glands common to some plant families such as the Apocynaceae and Orchidaceae).

The total essential oil content of plants is generally very low (often < 1%). However, many therapeutic oils are so potent that they are still active when used as herbal extracts or tinctures. Upon isolation, these oils are highly concentrated, and this is the form used widely by aromatherapists – mainly for external application, but sometimes diluted for internal consumption. Most oils consist of complex mixtures of chemical compounds, and it is often the unique chemical combination rather than a single component that is responsible for any therapeutic activity. The composition can vary according to the season, time of day, growing conditions and genetic make-up of the plant. Many oils contain over 100 individual compounds – these can generally be identified using gas chromatography coupled either with mass spectrometry (GC/MS) or flame ionization detector (GC/FID).

Chemotypes of oils

Essential oil composition can vary according to geographic and genetic factors, even though the same botanic species is involved – a phenomenon known as chemical polymorphism. When this occurs a terminology can be used where the Latin name is followed by the name of the chemical component most characteristic for that particular race of the plant (i.e. its chemotype, sometimes abbreviated to 'CT'), for example *Thymus vulgaris* linalol, *Thymus vulgaris* thymol.

Previously known as chemical forms, chemotypes are defined as 'those plants in a naturally occurring population which cannot be separated on morphological evidence, but which are readily distinguished by marked differences in the chemical composition of their essential oils' (Penfold and Willis, 1953). The term was later defined by Santesson in 1968 as 'chemically characterized parts of a population of morphologically indistinguishable individuals' (Keefover-Ring *et al.*, 2009). Each chemotype is genetically determined and highly heritable.

The significance of chemotypes can be appreciated in the case of tea tree oil. *Melaleuca alternifolia* has six or seven chemotypes, depending on the research source (Homer *et al.*, 2000; Keszei *et al.*, 2010). The Australian standard (ISO 4730:2017.E) for any oil traded as tea tree oil contains less than 10% 1,8-cineole, 30–48% terpinen-4-ol and 14–28% α-terpinene. For anyone intent on establishing a tea tree oil plantation, it is paramount the propagation material used is derived from plants with the correct chemotype. Hence to ensure that commercial plantations of *M. alternifolia* grown from seed are of the right chemotype, the chemotypes of tea tree oil from natural stands are genetically determined.

Major categories of aromatic oil compounds

Terpenoids
These are constructed from a series of isoprene units linked together in head-to-tail fashion, as described in Chapter 5, this volume. The most widely occurring terpenes are the smallest molecules, (i.e. the monoterpenes, $C_{10}H_{16}$) and their oxygenated derivatives such as ketones, aldehydes, alcohols, oxides, along with simple hydrocarbons. Their properties are determined by functional groups – oxygen-containing radicals attached to the carbon skeleton.

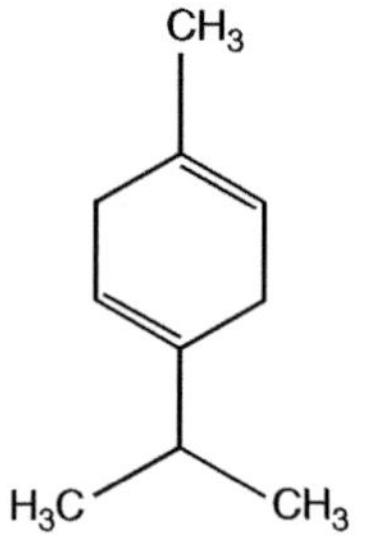

γ-terpinene – a monoterpene   menthol – a monoterpene
hydrocarbon   alcohol

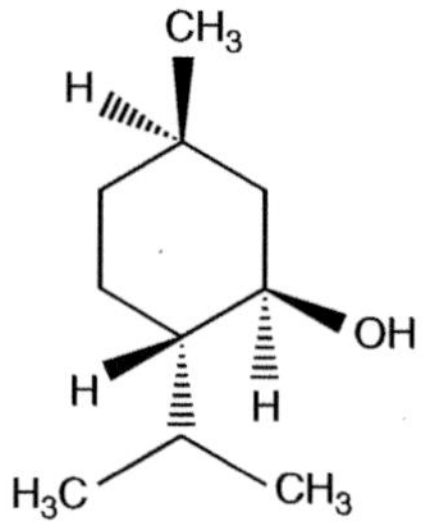

Sesquiterpenes ($C_{15}H_{24}$) and diterpenes ($C_{20}H_{32}$) may also occur in essential oils.

Phenylpropanoids
These compounds contain a benzene ring structure with an attached propane (C_3) side chain (see Chapter 2, this volume). The most common precursor is cinnamic acid, a derivative of the shikimic acid pathway. They include some aldehydes, phenols and phenolic ethers.

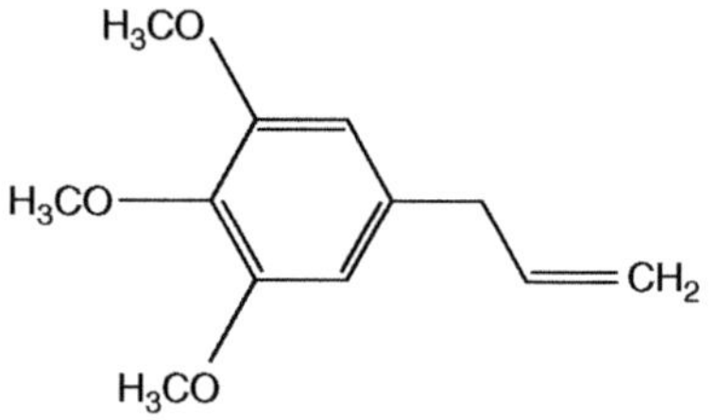

elemicin – a phenylpropanoid

Another major subclass consists of sulfur compounds, whose linear structures are non-terpenoid.

The subclasses or families of terpene essential oil constituents are listed in Table 8.1.

Table 8.1. Essential oil classification based on molecular class

Compound	Ending	Description	Compound	Essential oil example
Hydrocarbon	ene	Contains only carbon and hydrogen atoms ($C_{10}H_{16}$)	Pinene, limonene, terpenine	Pine, citrus
Alcohol	ol	Contains a hydroxyl group (OH) attached to the monoterpene skeleton	Menthol, terpenin-4-ol, geraniol, linalool	Peppermint, tea tree, lavender
Sesquiterpene	ene	Contains only carbon and hydrogen atoms ($C_{15}H_{24}$)	Chamazulene, β-caryophyllene, camaldulenic acid	Yarrow, cannabis, eucalyptus
Sesquiterpene alcohol	ol	Contains a hydroxyl group (OH) attached to the sesquiterpene skeleton	Bisabolol, santolol	German chamomile, sandalwood
Aldehyde	al	Terpenoids with a carbonyl group (C=O) and hydrogen bonded to a carbon	Citral (geranial/ neral), citronellal	Lemongrass, lemon myrtle, citronella
Cyclic aldehydes	hyde	Aldehyde group attached to a benzene ring	Cinnamic aldehyde, vanillin	Cinnamon, vanilla
Ketone	one	Contains a carbonyl group bonded to two carbon atoms	Camphor, thujone	Pennyroyal, thuja, Eucalyptus radiata
Phenol	ol	Hydroxyl group attached to a benzene ring	Thymol, eugenol, carvacrol	Thyme, clove, oregano
Phenolic ether	varies	Contains an oxygen between a carbon and benzene ring	Safrol, anethole, myristicin	Sassafras, aniseed, nutmeg
Oxide	ole	Has an oxygen bridging two or more carbons	Cineole, ascarldole	Eucalyptus, wormseed
Ester	ate	Condensation product of acid and alcohol	Methyl salicylate, linalyl acetate	Wintergreen, lavender
Lactone	in; one	Oxygenated lactone ring attached to a terpenoid skeleton	Coumarin, bergapten, umbelliferone	Angelica, bergamot, celery

The influence of chemical structures on essential oil therapeutics

Essential oils are readily absorbed into the body and across the blood–brain barrier because of their small molecular size and strong lipophilic nature. Compared to other compounds the therapeutic action of essential oils can be anticipated by knowledge of their chemistry – based primarily on the functional groups. As far back as 1937 the French master aromatherapist Rene-Maurice Gattefosse, referred to as the 'Father of Aromatherapy', developed a classified system for essential oils based on the chemistry of functional groups (Gattefosse, 1993). Hence, oils in the same molecular class are likely to exhibit similar therapeutic activities (see Table 8.2).

Monoterpene hydrocarbons

These are almost universal in essential oils, acting also as precursors of the more complex, oxidized terpenes. **Limonene**, found in citrus peel oils (up to 90%) and dill seed, *Anethum graveolens* (Apiaceae), is the precursor of menthol and carvone, major monoterpenoid constituents in mints, *Mentha* spp. (Lamiaceae). Limonene, *p*-**cymene** and other monoterpenes may also be formed as artefacts, due to the dehydration of terpene alcohols during drying or processing (Baser and Buchbauer, 2010).

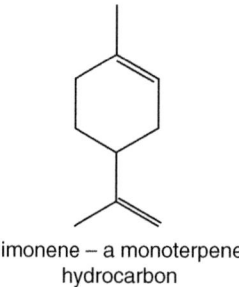

limonene – a monoterpene
hydrocarbon

Table 8.2. Properties of essential oil classes, based on functional group type

Compound class	Therapeutic actions
Hydrocarbons	Stimulant, decongestant, antiviral, antitumour
Alcohols	Antimicrobial, antiseptic, tonifying, spasmolytic
Sesquiterpene alcohols	Anti-inflammatory, anti-allergenic
Phenols	Antimicrobial, irritant, immune stimulating
Aldehydes	Spasmolytic, sedative, antiviral
Cyclic aldehydes	Spasmolytic
Ketones	Mucolytic, cell regenerating, neurotoxic
Esters	Spasmolytic, sedative, antifungal
Oxides	Expectorant, stimulant, antibacterial
Coumarins	UV sensitizing, antimicrobial, spasmolytic
Sesquiterpenes	Anti-inflammatory, antiviral
Phenylpropanes	Carminative, anaesthetic
Sesquiterpene lactones	Mucolytic, immune stimulating, anti-inflammatory

D-limonene is used clinically to dissolve cholesterol-containing gallstones, for heartburn and gastro-oesophageal reflux disease, while demonstrating low toxicity (Sun, 2007). D-limonene has performed well in a Phase 1 clinical trial, while oncological studies have revealed the monoterpene is chemopreventive against a range of cancer types (Jia *et al.*, 2012). Mechanisms contributing to these effects include stimulation of Phase 2 detoxifying enzymes responsible for removing carcinogens, and inhibition of hydroxymethylglutaryl-CoA (HMG-CoA) reductase (Franchomme, 2000; Schnaubelt, 2011). This rate-limiting enzyme of the mevalonate pathway is necessary for cholesterol metabolism, while it also promotes the proliferation of malignant cancer cells (Wong *et al.*, 2002).

Oil of turpentine (also known as terebinthina), derived from different species of pine (*Pinus* spp.), contains a high proportion of **α-** and **β-pinene**. To complicate things a little, each isomer has an enantiomer, so pinene may occur in four different forms, expressed as (+)-α-pinene, (–)-α-pinene, (+)-β-pinene and (–)-β-pinene. Pinene, structurally a bicyclic monoterpene, is derived from carbocarbation reactions with limonene, menthol or other monoterpene constituents (Baser and Buchbauer, 2010). In one study, the (+) enantiomers of pinene demonstrated antimicrobial activity, however, the (–) enantiomers showed no activity at all (da Silva *et al.*, 2012).

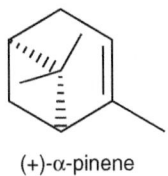

(+)-α-pinene

Turpentine oil is used as a rubefacient or liniment in rheumatic disease. Pinene has a pleasant aromatic odour and is an important component of many culinary spices, including black pepper (*Piper nigrum*, Piperaceae). **Myrcene** is a major component of hops (*Humulus lupus*, Cannabaceae) essential oil, and it occurs in many other species.

Recently there has been much focus on *Cannabis sativa* (Cannabaceae) chemovars (chemotypes) based on variability of cannabinoid and terpene profiles. While the hydrocarbons limonene, β-myrcene, and α-pinene (and two sesquiterpenoids) are present in all chemovars, other terpenes are restricted to specific chemovars (Mudge *et al.*, 2019). The terpenes provide fragrance differences between *C. sativa* varieties, and opinion is divided as to the significance of any therapeutic attributes due to synergism ('entourage effects') between cannabinoids and terpene constituents (Lewis *et al.*, 2018; Booth and Bohlmann, 2019).

Other widely distributed monoterpene hydrocarbons include **α-sabinene**, **terpinene**, *p*-cymene and **α-phellandrene**.

Alcohols
Alcohols have a hydroxyl group attached to a C_{10} hydrocarbon skeleton. Terpene alcohols are so highly valued for their fragrance, healing properties and gentle

reaction on skin and membranes that they have been termed 'friendly molecules' (Schnaubelt, 1989). Alcohols rank with phenols as being among the most potent antimicrobial essential oil compounds, but they do not contain the irritant properties of the latter.

Linalool (or linalol), an acyclic monoterpenoid alcohol, is enantiomeric – existing in two isoforms. The (*R*)-form (licareol) is found in rose, neroli and lavender oils, and is a major component of field mint (*Mentha arvensis*) and ho oil from *Cinnamomum camphora*, Lauraceae. The (*S*)-form (coriandrol) is found in oil of coriander seed (*Coriandrum sativum*, Apiaceae). The presence of more than 5% of the (*S*)-form in lavender oil (*Lavandula angustifolia*, Lamiaceae) is a strong indicator of adulteration (Leach, 2000).

S-(–)-linalool   R-(+)-linalool

enantiomeric forms of linalool

Linalool is widely distributed, being the major fragrance compound in many flowers, while also contributing to the flavour and biological activity of such diverse plants as grapevine leaves, lemon, basil, bergamot, tea leaves, thyme, cardamom and papaya fruit (Raguso and Pichersky, 1999; Duke and Beckstrom-Sternberg, 2001). It is also a major component of *Melaleuca ericifolia* (Myrtaceae).

Research into linalool-rich plants leaves from Brazil demonstrates potent effects on the central nervous system *in vivo*, including sedative, spasmolytic and hypothermic activity. Linalool has been shown to modulate glutamate activation expression *in vivo* and *in vitro* and may also inhibit GABAergic transmission (Elizabetsky *et al.*, 1999). Moreover, linalool produced an inhibitory effect on acetylcholine release and on the channel opening time of the neuromuscular junction, while demonstrating local anaesthetic action (Re *et al.*, 2000). Inhaled lavender oil and 3% linalool were shown to be anxiolytic in mice, without impairment of motor abilities (Linck *et al.*, 2009).

Tea tree oil is derived from leaves of *Melaleuca alternifolia* (Myrtaceae). To meet the Australian standard for tea tree oil (ISO 4730:2017.E) the **terpinen-4-ol** content should be in the range of 30–48%. Terpinen-4-ol is also a major ingredient in marjoram (*Origanum majorana*, Lamiaceae).

While tea tree oil is not the most powerful antimicrobial essential oil available, it is considered by some to be the ideal skin disinfectant, due to its activity against a wide range of microorganisms (including antibiotic resistant bacteria and fungi), its low incidence of irritation, anti-inflammatory action and ease of penetration (Altman, 1988; Carson *et al.*, 2006). Tea tree oil can be applied to most afflictions of the skin and orifices. The broad antimicrobial activity makes it useful for vaginal irritations since these can result from a variety of pathogenic organisms, including yeasts and bacteria (Williams and Home, 1995). Several studies have demonstrated that tea tree oil reduces lesion numbers in *Acne vulgaris*, equivalent to standard treatments including 5% benzoyl peroxide and 2% topical erythromycin, with low rate of adverse effects (Hammer, 2015). Antibacterial activity of tea tree oil has been attributed to increased membrane permeability induced by terpinen-4-ol, leading to leaking of potassium ions. Loss of intracellular ions disrupts cellular homoeostasis, inhibits respiration and metabolic processes within cells (Cox *et al.*, 2001; Carson *et al.*, 2006).

Melaleuca quinquenervia, the common coastal paperbark of Eastern Australia (also an invasive species in Florida USA), has two distinctive chemotypes. CT1 is comprised of the acyclic monoterpenoid alcohol *E*-**nerolidol** (74–95%), and linalool (14–30%). Nerolidol is closely related to **nerol** from neroli oil – derived from orange blossoms – and is similarly sweet in odour and gentle in action. Nerolidol itself has numerous documented activities, including antimicrobial, antifungal, antibiofilm, antiparasitic, anti-inflammatory, antinociceptive (pain relieving), antitumour, insect repellent and skin penetrating, making it an effective addition to topical medications (Chan *et al.*, 2016).

E-nerolidol

Peppermint oil is derived from the dried leaves and flowering tops of *Mentha piperita* (Lamiaceae). The oil consists of up to 50% menthol, the compound responsible for the cooling sensation induced by peppermint. Menthol triggers cold-sensitive thermoreceptors in the skin and mucosa (Jordt *et al.*, 2003). The taste and odour of peppermint oil are also influenced by some of its minor components, notably the menthol esters **jasmone** and **menthofuran**. The latter compound has a disagreeable odour and is mainly concentrated in young peppermint plants (Samuelsson, 1992). The ISO (International Organization for Standardization) standard for peppermint oil sets a maximum level of 8% for methofuran (de Groot and Schmidt, 2016). Peppermint is one of the best carminatives and the oil is sometimes administered in enteric-coated capsules for irritable bowel syndrome. Animal studies using peppermint oil demonstrated a significant spasmolytic effect, most likely as a result of the menthol content (Taddei *et al.*, 1988). Menthol is an ingredient in several pharmaceutical preparations and inhalants for congestion of the respiratory tract. Also it is used topically to relieve tension headaches. Antibacterial properties have also been demonstrated (Singh *et al.*, 2016), while peppermint extract inhibited biofilm formation of pathogenic bacteria and fungi (Sandasi *et al.*, 2011).

The so-called 'rose alcohols' include **geraniol** and **citronellol** from rose oil (*Rosa gallica*, Rosaceae), and scented geraniums (*Pelargonium* spp., Geraniaceae). **Nerol** is a stereoisomer of geraniol. The oil of wild bergamot (*Monarda fistulosa*, Lamiaceae) contains over 90% geraniol (Sell, 2010). Increasing dietary intake of geraniol was found to reduce the symptoms of irritable bowel syndrome in humans (Rizzelo *et al.*, 2018).

Perillyl alcohol, a metabolite derived from oxidation of limonene is found in *Perilla frutescens* (Lamiaceae) and a few other plants. The compound has undergone Phase II clinical trials at the National Cancer Institute for a series of cancer types including prostate, breast, ovarian and colorectal cancers and brain tumours (Buchbauer and Bohusch, 2010; Faria *et al.*, 2018). In one Phase II study, inhalation of perillyl alcohol was an effective therapeutic strategy capable of sustaining long-term regression (> 4 years survival) of recurrent glioma without significant side effects (da Fonseca *et al.*, 2013).

Borneol, a bicyclic alcohol found in rosemary essential oil (from *Salvia rosmarinus* syn. *Rosmarinus officinalis*, Lamiaceae) and many other essential oils, has potent analgesic and anti-inflammatory properties *in vivo* (Almeida *et al.*, 2013).

Sesquiterpene alcohols

Sesquiterpene alcohols are important fragrant compounds in a wide variety of species. The heartwood of East Indian and Australian sandalwood trees (*Santalum album* and *Santalum spicatum*, respectively, Santalaceae) produces an essential oil rich in sesquiterpenes, most notably **α-** and **β-santalol** (Howes *et al.*, 2004). Oil from the heartwood of *S. spicatum* may contain relatively high levels of ***E,E*-farnesol**. This sesquiterpene alcohol has a pleasant fragrance, however, it is also a suspected allergen. Hence cultivation of the species for essential oil production is focused on genetic strains with high santalol and low farnesol levels (Moniodis *et al.*, 2017).

Another sesquiterpene alcohol of note is **α-bisabolol**, an anti-inflammatory compound found in chamomile (*Matricaria recutita*, Asteraceae) and a minor component of sandalwood oil. Other actions associated with bisabolol are spasmolytic, antibacterial and wound healing, being partially responsible for the gentle action of chamomile. Being non-allergenic and non-irritant, the compound is a common ingredient in cosmetics, skin creams and baby-care products (Kamatou and Viljoen, 2010).

β-santolol

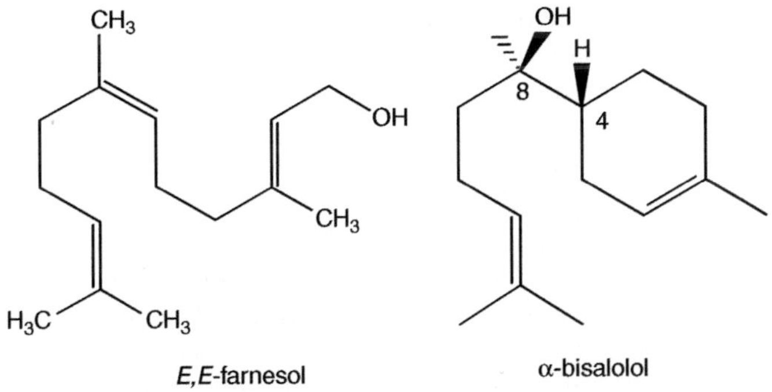

E,E-farnesol

α-bisalolol

Aldehydes

Aldehydes are highly reactive compounds in which one hydrogen atom is bonded to a carbonyl group at the end of a hydrocarbon chain. Monoterpene aldehydes such as those found in citrus oils correspond to their respective alcohol; note their names end in 'al', hence, for example, geraniol and citronellol are alcohols while geranial and citronellal are aldehydes.

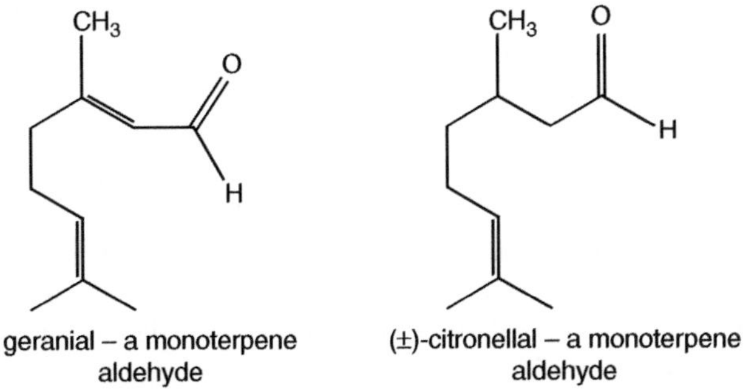

geranial – a monoterpene aldehyde

(±)-citronellal – a monoterpene aldehyde

Citrus essential oils are present in leaves, flowers and fruits of plants in the citrus family (Rutaceae), however, the main medicinal oils are found in the fruit peel. The best quality oils come from the bitter orange (*Citrus aurantium*) and the lemon (*Citrus limon*). Although the hydrocarbon limonene is a prominent constituent, the aroma of the oils is determined by the presence of aldehydes, namely the isomers **geranial** (citral A) and **neral** (citral B) – together known as **citral**. **Citronellal** is the dehydro analogue of citral.

Citral features as the dominant constituent in other citrus-flavoured oils such as those from lemon grass (*Cymbopogon citratus*, Poaceae). In a Malaysian study, the optimal time for harvesting leaves for maximum citral content was 6.7 months after planting out seedlings (Tajidin *et al.*, 2012). Other citral-rich plants include lemon balm (*Melissa officinalis*, Lamiaceae), lemon verbena (*Aloysia triphylla*, Verbenaceae) and the Australian lemon myrtle (*Backhousia citriodora*, Myrtaceae) which consists almost entirely of citral (Brophy *et al.*, 1995). The oil of the lemon-scented gum (*Corymbia citriodora*, Myrtaceae) consists mainly of citronellal with a small amount of the alcohol citronellol, the lemon-scented ironbark (*Eucalyptus staigeriana*, Myrtaceae) is rich in citral, while oil of lemon-scented tea tree (*Leptospermum petersonii*, Myrtaceae) contains both citral and citronellal.

Apart from its pleasant aroma, citral is valued for its sedative, antiviral and antimicrobial properties. Citral-rich oils derived from lemon grass and lemon-scented tea tree were shown to inhibit *Candida albicans* at more than four times the rate (zone of inhibition) of tea tree oil (Williams and Home, 1995). In one study *Backhousia citriodora* essential oil (95% citral) and a leaf paste were found to inhibit growth of bacteria and fungi, including human pathogens *Clostridium*, *Pseudomonas* and a hospital isolate of MRSA. Interestingly, the essential oil was slightly more potent than citral alone (Wilkinson *et al.*, 2003).

Many aldehydes are irritants, causing skin sensitivity in some people, somewhat restricting their use in topical applications. Franchomme (2000) found that application of limonene reduces the irritant action of citral and other aldehydes on skin.

Cyclic aldehydes

Also known as aromatic aldehydes, these are derived from shikimic acid and the terpene structure. They have characteristically sweet, pleasant odours and are found in some of our most well-known culinary herbs and spices, such as cinnamon and nutmeg.

Cinnamic aldehyde (syn. cinnamaldehyde), formed from reduction of cinnamic acid, is the characteristic flavour found in cinnamon and cassia barks (*Cinnamomum verum*, *Cinnamomum cassia*, Lauraceae). Cassia bark also contains coumarin – not present in true cinnamon bark – while cinnamon leaf is high in the closely related compound, eugenol (Senanayake and Wijesekera, 2004). Many of the pharmacological and therapeutic actions recorded for cinnamon bark and oil are due to cinnamaldehyde. These include anti-inflammatory, analgesic, antipyretic, diaphoretic, cardiotonic, antimicrobial and insecticidal. Cinnamon and cassia are classed as contact sensitizers due to the presence of the aldehydes, otherwise they are regarded as non-toxic (Vijayan and Thampuran, 2004).

The essential oil of cumin seed (*Cuminum cyminum*, Apiaceae) has up to 65% content of cuminaldehyde (van Wyk, 2013), while **benzaldehyde** is the main constituent of bitter almond (*Prunus dulcis* var. *amara*, Rosaceae) essential oil.

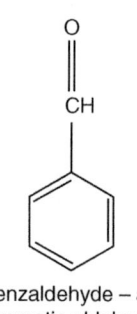

benzaldehyde – an
aromatic aldehyde

Phenolic essential oils

While present in only a relatively few aromatic herbs, phenolic volatile oils are among the most potent and potentially irritant compounds found in essential oils. Phenols are represented in both major classes of aromatic compounds – the monoterpenes and the phenylpropanoids. The major monoterpene phenols, **thymol** and its isomer **carvacrol**, are found in common thyme (*Thymus vulgaris*), oregano (*Origanum vulgare*) and bee balm (*Monarda didyma*) – all in the Lamiaceae family.

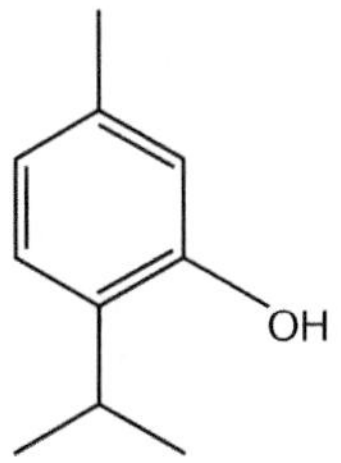

thymol – a monoterpene
phenol

The common garden thyme contains six essential oil chemotypes, being named after the dominant compounds as: thymol, linalool, geraniol, α-terpineol, **thujanol** and carvacrol (Thompson *et al.*, 2003). For medicinal use the thymol chemotype is preferred, however, other chemotypes are of interest to aromatherapists and gardeners.

Thymol is an expectorant, antimicrobial, anthelmintic and antispasmodic. Anti-inflammatory effects were demonstrated for thyme essential oil *in vitro*, however, the effect was attributed to the carvacrol content rather than thymol (Fachini-Queiroz *et al.*, 2012). Thymol is a dermal and mucous membrane irritant and caution is required in its use. The essential oil should never be ingested or applied undiluted to the skin.

Eugenol is a phenylpropanoid with a methoxyl and a hydroxyl group attached to the benzene ring. It is widely distributed in plants, one of the main sources being clove oil from flower buds of *Syzygium aromaticum* (Myrtaceae). Eugenol is a major constituent of bay leaf, tulsi, cinnamon leaf, nutmeg and allspice.

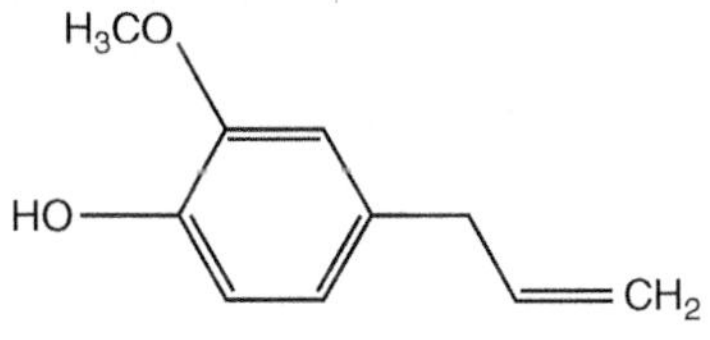

eugenol – a phenylpropanoid

Eugenol has antimicrobial properties akin to thymol, which, coupled with its anaesthetic properties, make it an effective disinfectant and cauterizing agent in dentistry (Valnet, 1980). Clove oil is an effective topical remedy for toothache, a powerful stimulant and aromatic. Clove powder is an essential ingredient in the 'Composition Powder' made famous by Samuel Thomson. Eugenol itself is a pharmacologically active essential oil compound; actions include antioxidant, neuroprotective, gastroprotective, hepatoprotective, spasmolytic and antimicrobial (Capasso *et al.*, 2000; Sayyah *et al.*, 2002; Pramod *et al.*, 2010).

Essential oil of sacred basil or tulsi (*Ocimum gratissimum*, Lamiaceae) contains three chemotypes dominated by eugenol, **methyl eugenol** and citral, respectively. The eugenol chemotype contains 43% eugenol and up to 32% of 1,8-cineole in one sample (Benitez *et al.*, 2009).

Phenolic ethers

Phenolic ethers – also known as alkenylbenzenes – are characterized by a phenylpropane (C_6C_3) skeleton with one or more ether functional groups attached to the benzene ring. There are two structural types, the first containing methoxyl groups similar to eugenol but with no free hydroxyl. Methyleugenol, **estragole**, **asarone** and **elemicin** are of this kind. In the second type, adjacent methoxyl groups form a methylenedioxyphenol ring containing two oxygen substitutes, as found in **safrole**, **myristicin** and **apiol**. **Piperine** from black pepper is also in this category, but with the addition of a heterocyclic nitrogen-containing ring, which also classifies it as an alkaloid.

Phenolic ethers are prominent constituents of common spices including black pepper, cloves, aniseed, nutmeg, celery seed, basil, parsley and tarragon. They are also found in some medicinal plants such as sweet flag (*Acorus calamus*, Acoraceae) – the main source of asarone – and cinnamon myrtle leaf (*Backhousia myrtifolia*, Myrtaceae), a rich source of elemicin and methyl eugenol.

estragole   safrole – a phenolic ether

Safrole is derived from the root bark of the sassafras tree (*Sassafras albidum*) and *Cinnamomum camphora*, both in the Lauraceae family. Safrole is also found as a minor constituent in cocoa, nutmeg and pepper, while sassafras bark was once the principle ingredient of 'root beer' (Hall, 1973). Use of this oil is restricted due to suspected carcinogenic properties – demonstrated in rodents following administration of high doses for long periods. No human toxicity has been reported.

Compounds such as safrole with methylenedioxyphenol structures also present a risk for herb–drug interactions. They are mechanism-based inhibitors of cytochrome P450 (CYP) enzymes, interacting with CYP haem iron to form intermediate complexes that produce DNA adducts. Safrole is the precursor used in the manufacture of MDMA (3,4-methylenedioxymethamphetamine, commonly known as ecstasy) (Gurley *et al.*, 2012; Gardner and McGuffin, 2013).

Myristicin (methoxysafrole) is found in nutmeg and mace (*Myristica fragrans*, Myristicaceae). It also occurs in black pepper, carrot, parsley and dill. Myristicin has similar safety concerns to safrol, being toxic in high doses (Al-Malahmeh *et al.*, 2017). Nutmeg itself exhibits narcotic and intoxicating properties, though it does have medicinal benefits in lower doses (Hall, 1973).

Methylchavicol also known as estragole is a major flavour constituent of basil (*Ocimum basilicum*, Lamiaceae). Studies of Turkish basil revealed the existence of seven chemotypes, including methyl chavicol and methyl eugenol chemotypes (Giachino *et al.*, 2014). As with other phenols, methyl chavicol is a skin irritant, though it is milder than methyl eugenol. Anethole, isomeric to methyl chavicol, is the familiar flavour characteristic in aniseed (*Pimpinella anisum*, Apiaceae), star anise (*Illicium verum*, Schisandraceae), fennel (*Foeniculum vulgare*, Apiaceae) and aniseed myrtle (*Anethola anisata*, Myrtaceae).

Ketones
Monoterpenoid ketones are cyclic compounds in which a carbonyl group is bonded to two carbon atoms. They are produced by oxidation of alcohols and are relatively stable molecules. They may be monocyclic as for **pulegone** or bicyclic like **camphor**.

pulegone – a monoterpenoid ketone

(–)-camphor – a bicyclic ketone

Ketones tend to be mucolytic, hence they are often inhaled in essential oil form for the relief of sinus and nasal congestion. **Isopinocamphone** is the main constituent of oil of hyssop (*Hyssopus officinalis*, Lamiaceae), imparting the typical mucolytic properties, while muscle relaxant activity has been demonstrated *in vivo* (Lu *et al.*, 2002). **Piperitone** is a constituent of the leaf oils of the so-called 'peppermint' group of the genus *Eucalyptus*. Both the broad- and narrow-leaved peppermints (*Eucalyptus dives* and *Eucalyptus radiata*, respectively) have piperitone-rich chemotypes (Boland *et al.*, 1991) which are employed as non-irritant, mucolytic agents for sinus congestion and bronchitis (Schnaubelt, 1995). The **verbenone** chemotype of rosemary essential oil, popular with aromatherapists, is used in a similar way.

The ketone **carvone** is optically isomeric. One isomer (+)-carvone is found in oil of caraway seed (*Carum carvi*, Apiaceae) and dill, whereas the other, (–)-carvone, is the main constituent of oil of spearmint (*Mentha spicata*, Lamiaceae). With the characteristic minty odour and taste, (–)-carvone is frequently used as an ingredient in toothpastes, mouthwashes and chewing gum. **Camphor** is derived from the heartwood of *Cinnamomum camphora*, in the (+) isomeric form. The (–) form occurs in feverfew (*Tanacetum parthenium*, Asteraceae), and in some lavender varieties. It is regarded as an undesirable constituent in true lavender (*Lavandula angustifolia*). Much of the camphor used in commerce is prepared synthetically from other monoterpenes. Camphor is a central-nervous-system stimulant, primarily used as a topical agent for its anti-pruritic, rubefacient and mucolytic properties. It is toxic in high doses.

Thujone was originally isolated from the arborvitae (*Thuja occidentalis*, Cupressaceae); however, it also occurs in some unrelated plants, particularly those of the *Artemisia* genus (Asteraceae) including wormwood (*Artemisia absinthium*) and mugwort (*Artemisia vulgaris*). Other sources are tansy (*Tanacetum vulgare*, Asteraceae), sage (*Salvia officinalis*, Lamiaceae) and clary sage (*Salvia sclarea*). Thujone is isomeric, the α and β forms reflecting different positions of the methyl attachment at C-4. Thujone is an alkylating agent, acting on proteins involved in neuronal signal conduction, and a GABA$_A$ receptor modulator, which can produce seizure-promoting effects, α-thujone being the more toxic (Gardner and McGuffin, 2013). **Pulegone**, a related ketone found in oil of pennyroyal (*Mentha pulegium*, Lamiaceae), has similar properties, and the same precautions for its use apply as for thujone-containing oils. The above herbs are classed as emmenagogues and potentially abortifacient, they are contraindicated for pregnancy.

Sage oil is the least toxic of this group despite containing up to 50% thujone, being a mixture of the two isomers. Distillation of young leaves has been found to minimize the thujone content (Schnaubelt, 2011).

Oxides

Monoterpene oxides are relatively reactive, unstable molecules in which an oxygen atom is situated between two carbons. In most cases the oxygen atom substitutes a carbon in the ring of a monoterpenoid – usually an alcohol. An example is **bisabolol oxide** from chamomile (*Matricaria chamomilla*, Asteraceae). However, in the two most significant oxides – **1,8-cineole** and **ascaridole** – found in essential oils, the oxygen atoms are attached outside the main hydrocarbon ring.

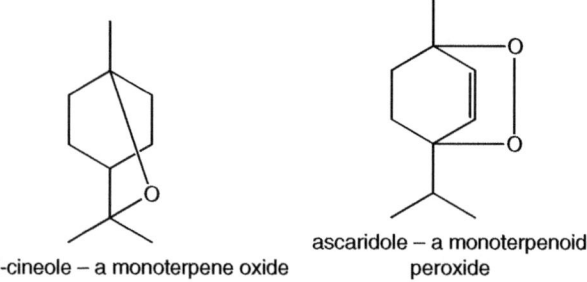

1,8-cineole – a monoterpene oxide   ascaridole – a monoterpenoid peroxide

1,8-Cineole (referred to also as cineole and eucalyptol) is one of the most widely distributed compounds in the plant world, being an oxidized derivative of other monoterpene compounds. It is a monoterpene ether, in which a secondary ring (containing the O) is attached to the first ring at positions 1 and 8. Cineole is the major constituent of eucalyptus oil derived from numerous species of *Eucalyptus* (Myrtaceae). It is also the major constituent of oil of cajuput (*Melaleuca cajuputi*, Myrtaceae) and of rosemary. Cineole is an expectorant and mucolytic agent, and a universal ingredient in cough lozenges and other respiratory medications.

Eucalyptus oils vary in aroma and quality according to the level of cineole, and the minor constituents present in each oil. Cineole-rich oils (from *Eucalyptus globulus* and *Eucalyptus polybractea*) are usually preferred for medicinal use where their expectorant property is highly valued; however, other species with a different balance of compounds are favoured by aromatherapists, such as ketone-containing 'peppermints' referred to above. All eucalyptus oils are renowned for their antiseptic and antimicrobial qualities. In one study of seven high-cineole-containing species, all were found to exhibit broad-spectrum antimicrobial action, however, there was significant variation in activity between species, most likely reflecting the contributions of minor constituents present in each essential oil, in addition to the 1,8-cineole content (Aldoghaim *et al.*, 2018). In another study, *E. globulus* essential oil (from leaf and fruit) was found to be the most effective inhibitor of MRSA and other drug-resistant infections of all the *Eucalyptus* species tested. It turned out essential oil from the fruits (large gum nuts) of *E. globulus* produced the maximum growth inhibition of the samples tested, though the fruits are relatively low in 1,8-cineole (Faleiro and Miguel, 2013). The synergistic activity of using two or more essential oils together can result in an even stronger antimicrobial activity, as demonstrated in an experiment where the addition of basil oil to eucalyptus oil increased the bactericidal activity 20-fold (Brud and Gora, 1989).

E. globulus leaf oil has shown a significant virucidal activity against influenza virus following exposure to oil vapours for only 10 minutes. In addition, this activity was observed without measurable adverse effect on the epithelial cell monolayers (Salehi *et al.*, 2019).

Ascaridole, an endoperoxide, has two oxygen atoms between two carbons. As the main constituent of wormseed oil from *Dysphania ambrosioides* formerly *Chenopodium ambrosioides* (Amaranthaceae) it is regarded as one of the most highly toxic of all essential oils, symptoms of toxicity including gastroenteritis, headache, facial flushing, vertigo and paraesthesia (Tisserand and Balacs, 1995; Monzote, 2007). Ascarisole is also a constituent of oil of boldo, from *Peumus boldo* (Monimaceae). *D. ambrosioides* is the source of 'apazote', a pungent garnish often used in Mexican food. A common weed in many parts of the world, there is a long history of its use in folk medicine in Central and South America, mainly as an anthelmintic and vermifuge (Monzote, 2007). It is mainly used in powder or tea form, which is relatively safe given the poor solubility of ascaridole. Monzote and his colleagues have demonstrated significant effects *in vivo* for *D. ambrosioides* essential oil against *Leishmania* spp. – protozoans that infect millions of people worldwide (Monzote *et al.*, 2007).

Esters

Esters are formed by reaction of terpene alcohols with acetic acid and other organic acids. They are among the most widespread volatile oil compounds – being found mainly in flowers – however, they are generally present in small amounts. Their distinctively fragrant aromas characterize many of the oils in which they appear. Lavender oil contains the alcohol linalool along with its ester **linalyl acetate**; the relative abundance of these two constituents is a marker for authenticity and quality. Bergamot oil from the peel of *Citrus bergamia* (Rutaceae) has a similar constituent profile to the true lavender oil, though with the addition of limonene.

linalyl acetate – a monoterpenoid acetate

Most esters are gentle, non-irritant compounds, whose action is mainly sedative and antispasmodic. Examples of these are also found in the oils of Roman chamomile (*Anthemis nobilis*, Asteraceae) and clary sage. Less benign esters are found in oils of wintergreen (*Gaultheria procumbens*, Ericaceae) and mustard (*Brassica nigra*, Brassicaceae) – see below.

Methyl salicylate is derived from oil of wintergreen. It is an aromatic ester derived from salicylic acid and methanol, though it can now be produced synthetically. Methyl salicylate is mainly used in topical applications and liniments as a counterirritant, antirheumatic, anti-inflammatory, analgesic and antipyretic. Internal administration is not recommended since it is quite toxic in large doses, and salicylates should be avoided by patients with any degree of kidney failure (Chin *et al.*, 2007).

Sulfur compounds

These are linear molecules containing one or more sulfur substituents. They are quite reactive. Most are based on the 'allyl' group structure ($CH_2=CHCH_2$), occurring mainly as aglycones to glucosinolates, covered in Chapter 4, this volume. **Allyl isothiocyanate** contains both nitrogen and sulfur and it is derived from oils of mustard (*Brassica* spp.) and horseradish (*Cochlearia armoracia*, Brassicaceae). Allyl isothiocyanate and similar compounds from the *Allium* genus are known to have antimicrobial and antitumour properties.

allyl isothiocyanate

Allyl sulfides are a group of compounds in which sulfhydryl groups (also known as thiols or mercaptans) are linked to the allyl skeleton. They are predominantly found in members of the *Allium* genus (Amaryllidaceae), notably in garlic (*Allium sativa*) and onion (*Allium cepa*). Their common precursor is the sulfur-containing amino acid cysteine. **Alliin** is the main sulfur compound found in raw garlic, however, it is quickly converted to **allicin** (diallyl thiosulfinate) upon processing or solvent extraction, and much alliin is lost in the process of dehydration (Block, 1985; Rahman, 2007).

Extraction techniques using different solvents have assisted in identification of oxidation and decomposition products of garlic and onions. These include: (i) **diallyl disulfide** – product of steam distillation of garlic; (ii) **propanthial S-oxide** (lachrymatory factor) – converted from a precursor by enzymic action when slicing onion; and (iii) **ajoene** – formed from decomposition of allicin. Several organosulfur compounds contribute to the distinctive garlic odour, including diallyl sulfide, allicin, **allyl mercaptan, ally methyl sulfide** and **disulfide** (Block, 1992; Rahman, 2007).

allicin – diallyl thisulfinate   (*E*)-ajoene

Most of the organosulfur compounds in garlic are antimicrobial, allicin being the most potent of all. The compound inhibits Gram-positive and Gram-negative bacteria at high dilutions, as well as *Candida* and other pathogenic fungi (Block, 1992). Mechanisms of antimicrobial action are thought to involve reaction with free thiol groups in enzymes such as cysteine proteases and alcohol dehydrogenases (Rahman, 2007).

Sulfur compounds in garlic also play a role in promotion of detoxification mechanisms, being modulators of both Phase I and Phase II metabolic enzymes. Allyl groups enhance gene transcription for CYP and other metabolizing enzymes, with diallyl trisulfide being the most potent (Tsai *et al.*, 2012). These actions block the biotransformation of nitrosamines and other carcinogens, adding to garlic's reputation as a chemopreventative agent (Rahman, 2007). Sulfur compounds, most notably allyl mercaptan, are histone deacetylase (HDAC) inhibitors, enabling them to reverse silent genes in cancer cells, resulting in cell cycle arrest and apoptosis (Nian *et al.*, 2008).

Cardioprotective actions of garlic have been demonstrated in animals and humans. Allyl derivatives such as the lipophilic ajoene have been shown to moderate many risk factors for cardiovascular diseases. Their actions include vasodilatory, antithrombotic, platelet aggregation inhibition, antiatherogenic, antihypertensive, antihyperlipidaemic, hypoglycaemic, and anti-inflammatory (Rahman, 2007; Tsai *et al.*, 2012).

References

Aldoghaim, F.S., Flematti , G.R. and Hammer, K.A. (2018) Antimicrobial activity of several cineole-rich Western Australian *Eucalyptus* essential oils. *Microorganisms* 6(4), 122. doi: 10.3390/microorganisms6040122

Al-Malahmeh, A.J., Al-Ajlouni, A., Wesseling, S., Soffers, A.E., Al-Subeihi, A., *et al.* (2017) Physiologically based kinetic modeling of the bioactivation of myristicin. *Archives of Toxicology* 91(2), 713–734. doi: 10.1007/s00204-016-1752-5

Almeida, J.R., Souza, G.R., Silva, J.C., Saraiva, S.R., de Olveira, R., *et al.* (2013) Borneol, a bicyclic monoterpene alcohol, reduces nociceptive behavior and inflammatory response in mice. *The Scientific World Journal* 2013, 808460. doi: 10.1155/2013/808460

Altman, P. (1988) Australian tea tree oil. *Current Drug Information* 2, 62–64.

Assar, N. (2007) Le nez – the nose. *The Fafai Journal* 9(10), 59–60.

Baser, K.H.C. and Buchbauer, G. (eds) (2010) *Handbook of Essential Oils: Science, Technology, and Applications*. CRC Press, Boca Raton, Florida.

Benitez, N.P., Meléndez-León, E.M. and Stashenko, E.E. (2009) Eugenol and methyl eugenol chemotypes of essential oil of species *Ocimum gratissimum* L. and *Ocimum campechianum* Mill. from Colombia. *Journal of Chromatographic Science* 47(9), 800–803. https://doi.org/10.1093/chromsci/47.9.800

Block, E. (1985) The chemistry of garlic and onions. *Scientific American* 252(3), 94–99.

Block, E. (1992) The organosulfur chemistry of the genus *Allium* – implications for the organic chemistry of sulfur. *Angewandte Chemie International Edition* 31(9), 135–1178. https://doi.org/10.1002/anie.199211351

Boland, D.J., Brophy, J.J. and House, A.P.N. (1991) *Eucalyptus Leaf Oils*. Intaka Press, Melbourne, Victoria, Australia.

Booth, J.K. and Bohlmann, J. (2019) Terpenes in *Cannabis sativa* – from plant genome to humans. *Plant Science* 284, 67–72. https://doi.org/10.1016/j.plantsci.2019.03.022

Brophy, J.J., Goldsack, R.J., Fookes, C.J.R. and Forster, P.I. (1995) Leaf oils of the genus *Backhousia* (Myrtaceae). *Journal of Essential Oil Research* 7, 237–254. https://doi.org/10.1080/10412905.1995.9698514

Brud, W.S. and Gora, J. (1989) Biological activity of essential oils and its possible applications. In: Bhattacharyya, S., Sen, N. and Sethi, K.L. (eds) *Proceedings of the 11th International Congress on Essential Oils, Fragrances and Flavours, 12–16 November 1989, New Delhi, India*, Vol. 2. CAB International, Wallingford, UK, pp. 13–23.

Buchbauer, G. and Bohusch, R. (2010) Biological activities of essential oils. In: Baser, K.H.C. and Buchbauer, G. (eds) *Handbook of Essential Oils: Science, Technology, and Applications*. CRC Press, Boca Raton, Florida, pp. 281–321.

Capasso, R., Pinto, L., Vuotto, M.L. and Di Carlo, G. (2000) Preventative effect of eugenol on PAF and ethanol-induced gastric mucosal damage. *Fitoterapia* 71(S1), 131–137. https://doi.org/10.1016/S0367-326X(00)00188-X

Carson, C.F., Hammer, K.A. and Riley, T.V. (2006) *Melaleuca alternifolia* (tea tree) oil: a review of antimicrobial and other medicinal properties. *Clinical Microbiology Reviews* 19(1), 50–62.

Chan, W.K., Tan, L.T., Chan, K.G., Lee, L.H. and Goh, B.H. (2016) Nerolidol: a sesquiterpene alcohol with multi-faceted pharmacological and biological activities. *Molecules* 21(5), 529. doi: 10.3390/molecules21050529

Chin, R.L., Olson, K.R. and Dempsey, D. (2007) Salicylate toxicity from ingestion and continued dermal absorption. *The California Journal of Emergency Medicine* 8(1), 23–25. Available at: https://www.ncbi.nlm.nih.gov/pmc/articles/PMC2859737/ (accessed 5 December 2020).

Cox, S., Mann, C.M., Markham, J., Gustafson, J.E., Warmington, J.R. and Wyllie, S.G. (2001) Determining the antimicrobial actions of tea tree oil. *Molecules* 6(2), 87–91. doi: 10.3390/60100087

da Fonseca, C.O., Teixeira, R.M., Silva, J.C.T., Fischer, J.D., Meirelles, O.C., *et al.* (2013) Long-term outcome in patients with recurrent malignant glioma treated with per-illyl alcohol inhalation. *Anticancer Research* 33(12), 5625–5631. Available at: http://ar.iiarjournals.org/content/33/12/5625.long (accessed 5 December 2020).

da Silva, A.C., Lopes, P.M., Azevedo, M.M., Costa, D.C., Alviano, C.S. and Alviano, D.S. (2012) Biological activities of α-pinene and β-pinene enantiomers. *Molecules* 17(6), 6305–6316. doi: 10.3390/molecules17066305

de Groot, A.C. and Schmidt, E. (2016) *Essential Oils: Allergy Contact and Chemical Composition*. CRC Press, Boca Raton, Florida.

Duke, J.A. and Beckstrom-Sternberg, S.M. (2001) *Dr. Duke's Phytochemical and Ethnobotanical Databases*. National Germplasm Research Laboratory, Agricultural Research Studies, United States Department of Agriculture (USDA), Beltsville, Maryland. Available at: http://dx.doi.org/10.15482/USDA.ADC/1239279 (accessed 6 December 2020).

Elizabetsky, E., Brum, L.F. and Souza, D.O. (1999) Anticonvulsant properties of linalool in glutamate-related seizure models. *Phytomedicine* 6(2), 107–113. doi: 10.1016/s0944-7113(99)80044-0

Fachini-Queiroz, F.C., Kummer, R., Estevão-Silva, C.F., Carvalho, M.D., Cunha, J.M., *et al.* (2012) Effects of thymol and carvacrol, constituents of *Thymus vulgaris* L. essential oil, on the inflammatory response. *Evidence-based Complementary and Alternative Medicine: eCAM* 2012: 657026. doi: 10.1155/2012/657026

Faleiro, M.L. and Miguel, M.G. (2013) Use of essential oils and their components against multidrug-resistant bacteria. In: Rai, M. and Kon, K. (eds) *Fighting Multidrug Resistance with Herbal Extracts, Essential Oils and their Components*. Academic Press, Cambridge, Massachusetts, pp.65–94.

Faria, G., Silva, E., Da Fonseca, C. and Quirico-Santos, T. (2018) Circulating cell-free DNA as a prognostic and molecular marker for patients with brain tumors under perillyl alcohol-based therapy. *International Journal of Molecular Sciences* 19(6), 1610. doi: 10.3390/ijms19061610

Franchomme, P. (2000) Essential oils and cancer. In: Schnaubelt, K. (ed.) *Proceedings of the 4th Scientific Wholistic Aromatherapy Conference, 10–12 November 2000, San Francisco*. Pacific Institute of Aromatherapy, San Rafael, California.

Gardner, Z. and McGuffin, M. (2013) *Botanical Safety Handbook*, 2nd edn. American Herbal Products Association (AHPA), CRC Press, Boca Raton, Florida.

Gattefosse, R.M. (1993) *Gattefosse's Aromatherapy*. Translated by R. Tisserand and L. Davies. CW Daniel Co Ltd, Saffron-Walden, UK.

Giachino, R.R.A., Sönmez, C., Tonk, F.A., Bayram, E., Yüce, S., *et al.* (2014) RAPD and essential oil characterization of Turkish basil (*Ocimum basilicum* L.). *Plant Systemics and Evolution* 300, 1779–1791. https://doi.org/10.1007/s00606-014-1005-0

Guba, R. (2018) The modern alchemy of carbon dioxide extraction. *Aromatherapy Today* Special edition, August, 35–39.

Gurley, B.J., Fifer, E.K. and Gardner, Z. (2012) Pharmacokinetic herb–drug interactions (Part 2). *Planta Medica* 78(13), 1490–1514. doi: 10.1055/s-0031-1298331

Hall, R. (1973) Toxicants occurring naturally in spices and flavours. In: National Research Council (ed.) *Toxicants Occurring Naturally in Foods*. The National Academies Press, Washington, DC, pp. 448–463. https://doi.org/10.17226/21278

Hammer, K.A. (2015) Treatment of acne with tea tree oil (*Melaleuca*) products: a review of efficacy, tolerability and potential modes of action. *International Journal of Antimicrobial Agents* 45(2), 106–110. doi: 10.1016/j.ijantimicag.2014.10.011

Homer, L.E., Leach, D.N., Lea, D., Lee, L.S., Henry, R.J. and Baverstock, P.R. (2000) Natural variation in the essential oil content of *Melaleuca alternifolia* Cheel (Myrtaceae). *Biochemical Systematics and Ecology* 28(4), 67–382. https://doi.org/10.1016/S0305-1978(99)00071-X

Howes, M.R., Simmonds, M.S.J. and Kite, G.C. (2004) Evaluation of the quality of sandalwood essential oils by gas chromatography–mass spectrometry. *Journal of Chromatography A* 1028(2), 307–312. https://doi.org/10.1016/j.chroma.2003.11.093

Jia, S., Peng, G., Zhangi, M., Chen, Y., Lei, B., *et al.* (2012) Induction of apoptosis by D-limonene is mediated by inactivation of Akt in LS174T human colon cancer cells. *Oncology Reports* 29(1), 349–354. doi: 10.3892/or.2012.2093

Jordt, S., McKemy, D.D. and Julius, D. (2003) Lessons from peppers and peppermint: the molecular logic of thermosensation. *Current Opinion in Neurobiology* 13(4), 487–492. https://doi.org/10.1016/S0959-4388(03)00101-6

Kamatou, G.P.P. and Viljoen, A.M. (2010) A review of the application and pharmacological properties of α-bisabolol and α-bisabolol-rich oils. *Journal of the American Oil Chemistry Society* 87, 1–7. https://doi.org/10.1007/s11746-009-1483-3

Keefover-Ring, K., Thompson, J.D. and Linhart, Y.B. (2009) Beyond six scents: defining a seventh *Thymus vulgaris* chemotype new to southern France by ethanol extraction. *Flavour and Fragrance Journal* 24(3), 117–122. https://doi.org/10.1002/ffj.1921

Keszei, A., Hassan, Y. and Foley, W.J. (2010) A biochemical interpretation of terpene chemotypes in *Melaleuca alternifolia*. *Journal of Chemical Ecology* 36(6), 652–661. doi: 10.1007/s10886-010-9798-y

Leach, D. (2000) *Testing for Quality of Lavender Oils*. Lavender 2000 Conference. Australian Lavender Industry, Wagga Wagga, New South Wales, Australia.

Lewis, M.A., Russo, E.B. and Smith, K.M. (2018) Pharmacological foundations of *Cannabis* chemovars. *Planta Medica* 84(4), 225–233. doi: 10.1055/s-0043-122240

Linck, V.M., Silva, L.A., Figueiro, M., Piato, A.L., Herrmann, A.P., *et al.* (2009) Inhaled linalool-induced sedation in mice. *Phytomedicine* 16(4), 303–307. doi: 10.1016/j.phymed.2008.08.001

Lu, M., Battinelli, L., Daniele, C., Melchioni, C., Salvatore, G. and Mazzanti, G. (2002) Muscle relaxing activity of *Hyssopus officinalis* essential oil on isolated intestinal preparations. *Planta Medica* 68(3), 213–216. doi: 10.1055/s-2002-2313

Malle, B. and Schmickl, H. (2012) *The Essential Oil Maker's Handbook*. Spikehorn Press, Austin, Texas.

Moniodis, J., Jones, C.G., Renton, M., Plummer, J.A., Barbour, E.L., *et al.* (2017) Sesquiterpene variation in West Australian sandalwood (*Santalum spicatum*). *Molecules* 22(6), 940. doi: 10.3390/molecules22060940

Monzote, L. (2007) Essential oil from *Chenopodium ambrosioides*: a return to traditional medicine? *International Journal of Essential Oil Therapeutics* 1(1), 179–183.

Monzote, L., Montalvo, A.M., Scull, R., Miranda, M. and Abreu, J.C. (2007) Activity, toxicity and analysis of resistance of essential oil from *Chenopodium ambrosioides* after intraperitoneal, oral and intralesional administration in BALB/c mice infected with *Leishmania amazonensis*: a preliminary study. *Biomedicine and Pharmacotherapy* 61(2–3), 148–153. https://doi.org/10.1016/j.biopha.2006.12.001

Mudge, E.M., Brown, P.N. and Murch, S.J. (2019) The terroir of *Cannabis*: terpene metabolomics as a tool to understand *Cannabis sativa* selections. *Planta Medica* 85(09–10), 781–796. doi: 10.1055/a-0915-2550

Nian, H., Delage, B., Pinto, J.T. and Dashwood, R.H. (2008) Allyl mercaptan, a garlic-derived organosulfur compound, inhibits histone deacetylase and enhances Sp3 binding on the P21WAF1 promoter. *Carcinogenesis* 29(9), 1816–1824. doi: 10.1093/carcin/bgn165

Pavia, D., Lampman, G. and Kriz, G. (1982) *Introduction to Organic Laboratory Techniques.* Saunders College Publishing, Philadelphia, Pennsylvania.

Penfold, A.R. and Willis, J.L. (1953) Physiological forms in *Eucalyptus citriodora*, Hooker. *Nature* 171(4359), 883–884. doi: 10.1038/171883b0

Pramod, K., Ansari, S.H. and Ali, J. (2010) Eugenol: a natural compound with versatile pharmacological actions. *Natural Product Communications* 5(12), 1999–2006.

Raguso, R.A. and Pichersky, E. (1999) New perspectives in pollination biology: floral fragrances. A day in the life of a linalool molecule: chemical communication in a plant-pollinator system. Part 1: linalool biosynthesis in flowering plants. *Plant Species Biology* 14(2), 95–120. https://doi.org/10.1046/j.1442-1984.1999.00014.x

Rahman, M.S. (2007) Allicin and other functional active components in garlic: health benefits and bioavailability. *International Journal of Food Properties* 10(2), 245–268. https://doi.org/10.1080/10942910601113327

Re, L., Barocci, S., Sonnino, S., Mencarelli, A., Vivana, C., *et al.* (2000) Linalool modifies the nicotinic receptor–ion channel kinetics at the mouse neuromuscular junction. *Pharmacological Research* 42(2), 177–182. https://doi.org/10.1006/phrs.2000.0671

Rizzelo, F., Campieri, M., Comparone, A., De Fazio, L., Candela, M., *et al.* (2018) Dietary geraniol ameliorates intestinal dysbiosis and relieves symptoms in irritable bowel syndrome patients: a pilot study. *BMC Complementary and Alternative Medicine* 18(1), 338. doi: 10.1186/s12906-018-2403-6

Salehi, B., Sharifi-Rad, J., Quispe, C., Llaique, H., Villalobos, M., *et al.* (2019) Insights into *Eucalyptus* genus chemical constituents, biological activities and health-promoting effects. *Trends in Food Science & Technology* 91, 609–624. https://doi.org/10.1016/j.tifs.2019.08.003.

Samuelsson, G. (1992) *Drugs of Natural Origin.* Swedish Pharmaceutical Press, Stockholm.

Sandasi, M., Leonard, C.M., Van Vuuren, S.F. and Viljoen, A.M. (2011) Peppermint (*Mentha piperita*) inhibits microbial biofilms *in vitro. South African Journal of Botany* 77(1), 80–85. https://doi.org/10.1016/j.sajb.2010.05.011

Sayyah, M., Valizadeh, J. and Kamalinejad, M. (2002) Anticonvulsant activity of the leaf essential oil of *Laurus nobilis* against pentylenetetrazole- and maximal electroshock-induced seizures. *Phytomedicine* 9(3), 212–216. https://doi.org/10.1078/0944-7113-00113

Schnaubelt, K. (1989) Friendly molecules. *International Journal of Aromatherapy* 2(2), 20–22.

Schnaubelt, K. (1995) *Advanced Aromatherapy.* Healing Arts Press, Rochester, Vermont.

Schnaubelt, K. (2011) *The Healing Intelligence of Essential Oils.* Healing Arts Press, Rochester, Vermont.

Sell, C. (2010) Chemistry of essential oils. In: Baser, K.H.C. and Buchbauer, G. (eds) *Handbook of Essential Oils. Science, Technology and Applications*. CRC Press, Boca Raton, Florida, pp. 121–150.

Senanayake, U.M. and Wijesekera, R.O.B. (2004) Chemistry of cinnamon and cassia. In: Ravindran, P.N., Babu, N. and Shylaja, M. (eds) *Cinnamon and Cassia. The Genus* Cinnamomum. CRC Press, Boca Raton, Florida, pp. 80–120.

Singh, R., Shushni, M.A.M. and Belkheir, A. (2016) Antibacterial and antioxidant activities of *Mentha piperita* L. *Arabian Journal of Chemistry* 8(3), 322–328. https://doi.org/10.1016/j.arabjc.2011.01.019

Sun, J. (2007) D-limonene: safety and clinical applications. *Alternative Medicine Review* 12(3), 259–264. Available at: http://archive.foundationalmedicinereview.com/publications/12/3/259.pdf (accessed 6 December 2020).

Taddei, I., Giachetti, D., Taddei, E. and Mantovani, P. (1988) Spasmolytic activity of peppermint, sage, and rosemary essences and their major constituents. *Fitoterapia* LIX, 463–468.

Tajidin, N.E., Ahmad, S.H., Rosenani, A.B., Azimah, H. and Munirah, M. (2012) Chemical composition and citral content in lemongrass (*Cymbopogon citratus*) essential oil at three maturity stages. *African Journal of Biotechnology* 11(11), 2685–2693. Available at: https://academicjournals.org/article/article1380815053_Tajidin%20et%20al.pdf (accessed 14 June 2012).

Thompson, J.D., Chalchat, J.-C., Michet, A., Linhart, Y.B. and Ehlers, B. (2003) Qualitative and quantitative variation in monoterpene co-occurrence and composition in the essential oil of *Thymus vulgaris* chemotypes. *Journal of Chemical Ecology* 29, 859–880. https://doi.org/10.1023/A:1022927615442

Tisserand, R. and Balacs, T. (1995) *Essential Oil Safety*. Churchill Livingstone, Edinburgh, UK.

Tsai, C.-W., Chen, H.-W., Sheen, L.-Y. and Lii, C.-K. (2012) Garlic: health benefits and actions. *Biomedicine* 2(1), 17–29. https://doi.org/10.1016/j.biomed.2011.12.002

Valnet, J. (1980) *The Practice of Aromatherapy*. Destiny Books, New York.

van Wyk, B.E. (2013) *Culinary Herbs and Spices of the World*. University of Chicago Press, Chicago, Illinois.

Vijayan, K.K. and Thampuran, R.V.A. (2004) Pharmacology and toxicology of cinnamon and cassia. In: Ravindran, P.N., Babu, N. and Shylaja, M. (eds) *Cinnamon and Cassia. The Genus* Cinnamomum. CRC Press, Boca Raton, Florida, pp. 259–284.

Wilkinson, J.M., Hipwell, M., Ryan, T. and Cavanagh, H.M.A. (2003) Bioactivity of *Backhousia citriodora*: antibacterial and antifungal activity. *Journal of Agricultural and Food Chemistry* 51(1), 76–81. https://doi.org/10.1021/jf0258003

Williams, L. and Home, V. (1995) A comparative study of some essential oils for potential use in topical applications for the treatment of the yeast *Candida albicans*. *Australian Journal of Medical Herbalism* 7(3), 57–62. Available at: https://www.cabi.org/ISC/abstract/19960304327 (accessed 6 December 2020).

Wong, W.W.-L., Dimitroulakos, J., Minden, M.D. and Penn, L.Z. (2002) HMG-CoA reductase inhibitors and the malignant cell: the statin family of drugs as triggers of tumor-specific apoptosis. *Leukemia* 16(4), 508–519. https://doi.org/10.1038/sj.leu.2402476

Wrigley, J. and Fagg, M. (1990) *Aromatic Plants*. Angus and Robertson, Sydney, New South Wales, Australia.

Zozulza, S., Echeverri, F. and Nguyen, T. (2001) The human olfactory repertoire. *Genome Biology* 2(6), 1–12. doi: 10.1186/gb-2001-2-6-research

Polysaccharides

9

Introduction

Polysaccharides are high-molecular-weight polymers consisting of chains of sugars (monosaccharides or oligosaccharides) with chemical linkages. The simplest polysaccharides are starch and cellulose, which are polymers of glucose only, arranged in linear shape and connected by glycosidic linkages, α-linkages for starch and β-linkages for cellulose. Starch is a combination of two polymers, amylose, a linear (1→4)-α-D-glucan, and amylopectin composed of linear chains of glucose residues with α-(1,6) linkages. Humans are equipped with enzymes to break down α-linkages but not β-linkages, hence starch provides energy while cellulose provides dietary fibre.

Polysaccharides are difficult to represent in the usual structural formulas as they are so big – and often the identity and number of all the sugars present is unknown. Usually the chemical structures are represented by a section of the molecule, or by a series of named sugars with indications of the bonding arrangement and branching patterns – if any.

Polysaccharides are universal in the plant and fungal kingdoms. Their functions include food storage, protection of membranes, and maintaining rigidity of cell walls in plants and fungi, whereas for seaweeds they help maintain the flexibility required for life in the ocean (Bruneton, 1995). Examples of carbohydrate polymers and their significance to plants and humans are shown in Table. 9.1.

Polysaccharides are insoluble in organic solvents – they precipitate in alcohol. Herbal tinctures, which are made using alcoholic solvents of 45% strength or higher, are therefore of little use for polysaccharide extraction. The degree of water solubility depends on the polysaccharide structure. Linear polymers (mucilages) are less water soluble and tend to precipitate at high temperatures and form viscous or slimy solutions. Branched polymers (gums) are more water soluble and form gels, often referred to as 'gummy' or 'sticky'.

DOI: 10.1079/9781789243079.0009

Table 9.1. Carbohydrate polymers: sources and significance

Polymer type	Sugars – molecular type	Significance to plants	Significance to humans
Starch	Amylose, amylopectin	Main food reserve	Energy source Elevates blood glucose
Inulin	Fructofuranase	Alternative food reserve	Prebiotic, immune stimulant
Cellulose	Glucopyranose	Plant cell wall constituent, structural component	Pharmaceutical – tablet fillers, coatings
Pectin	Galactosyluronic residues	Fruit cell wall constituents, assists development of plant cells	Setting agents for jam and jellies Vehicle for nasal, ocular and oral drug delivery
Alginic acid	D-mannuronic and L-glucuronic acid residues	Brown algae: helps maintain flexibility of extracellular wall matrix and regulates plant growth. Commercially extracted from *Laminaria* and *Macrocystis* spp.	Stabilizing and thickening agents for food industry Surgical dressings Promotes wound healing
Agar	Sulfated agarose and agaropectin	Red algae: helps maintain flexibility of cell walls. Commercially extracted from *Gracilaria* and *Gelidium* spp.	Biology laboratories – bacterial culture media Gelling agent in foods
Carrageenan	Sulfated D-galactose and 3,6-anhydro-D-galactose	Red algae: helps maintain flexibility of cell walls. Obtained commercially from Irish moss (*Chondrus crispus*), *Kappaphycus* and *Betaphycus* spp.	Production of hydrogels for pharmaceuticals Cough lozenges Food additive (yoghurt, custards, chocolate milk)
Fucoidan	Sulfated L-fucose, D-galactose and various combinations of monosaccharides	Protects brown algae against dehydration and oxidation. Extracted from *Undaria pinnatifida*, *Fucus vesiculosus* and other species	Antitumour and anti-inflammatory agent
Chitin	N-acetylglucosamine	Fungal cell wall constituents, shells of crustaceans	Source of chitosan, chelating agent used in water purification and wound-healing preparations

Tragacanthe	D-galacturonic acid, D-galactose, L-fucose, L-arabinose and D-xylose	Dried gum from roots of *Astragalus gummifera* and other species	Suspension and binding agent for tablets
Gum arabic, gum acacia	L-arabinose, D-galactose, glucuronic acid L-rhamnose, in a 3:3:1:1 ratio (for gum arabic)	Dried gum from stems and bark of *Senegalia* and *Acacia* spp. Used to seal small wounds or infections	Suspension and binders for tablets Antidiarrhoeal agents Emulsifiers for essential oils
Kayara gum	L-rhamnose, D-galactose, and with a high proportion of D-galacturonic acid and D-glucuronic acid residues	Dried gum from trunks of the tree *Sterculia urens* and related species	Bulk laxative and suspension agent Lozenges Denture powders
Seed gums (legumes)	Gallactomannans with variable galactose to mannose ratios	Dried gum from seeds of carob or locust bean (*Ceratonia siliqua*), guar gum (*Cyamopsis tetragonoloba*) and other Fabaceae	Stabilizing, thickening agent, edible coating for food industry Gluten substitute for baking Cholesterol lowering
Xanthan gum	Glucose, D-mannose, glucuronic acid in a 2:2:1 ratio	Produced industrially from fermentation of glucose by the bacteria *Xanthomonas campestris*	Stabilizing, thickening agent in food and cosmetic industries Vehicle for drug delivery
Psyllium seed and hull	Arabinoxylan with xylose and arabinose in a 3:1 ratio	Mucilage-containing seeds of *Plantago afra* (syn. *P. psyllium*) and *Plantago ovata*. Swells in water to form a gel	Bulk laxative Demulcent Cholesterol lowering

Gums

Many plants produce gummy exudates when the bark is damaged by injury or fungal attack, and which serves to heal the wound. The exudate often dries to a hard amorphous mass, being produced in sufficient abundance by some species of trees and shrubs to warrant collection and commercial utilization. Gum exudates are readily obtainable in relatively pure, undegraded form though in some cases purification is still required. These polysaccharides often occur in association with a protein.

Gums are made up of branching chains of chemically linked sugars (monosaccharides) or their salts, and uronic acid derivatives (uronic acids are oxidation products of sugars). They may have acidic, basic or neutral characteristics. They are usually heterogeneous – composed of various monosaccharide residues and uronic acids.

Seed gums are obtained from certain types of seed endosperms. They are common in the legume (Fabaceae) family, for example locust bean gum (*Ceratonia siliqua*) and guar gum (*Cyamopsis tetragonoloba*) (Bourbon *et al.*, 2010; Barak and Mudgil, 2014). Xanthan gum is not a plant-based substance; it is produced by bacterial fermentation of carbohydrates, most notably glucose (Faria *et al.*, 2011; Petri, 2015).

Gums are widely used in the pharmaceutical industry, for example **gum tragacanthe** from *Astragalus* spp. (Fabaceae), having several advantages over synthetic polymers. These include being edible, safe, biodegradable, low cost, readily available and environmentally friendly (Goswami and Naik, 2014).

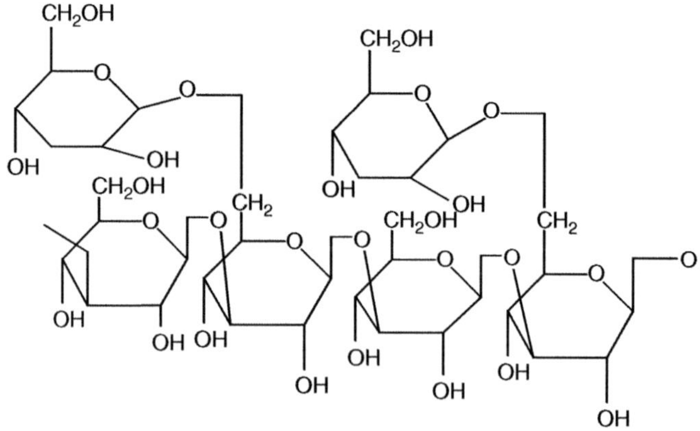

astragaloglucan – branch-chained
polysaccharide from *Astragalus* spp.

Gum arabic

On the basis of recent molecular and phylogenetic research, taxonomists have split the *Acacia* genus (Fabaceae family) into five separate genera, with only the Australian species (of which there are over 1000) retaining the name *Acacia*. In Africa, the previously named *Acacia* species are now divided into two genera, *Senegalia* and *Vachellia* (Kyalangalilwa *et al.*, 2013). Hence **gum arabic** is now derived from *Senegalia senegal* (previously *Acacia senegal*) and the closely related species *Senegalia seyal*. Phytochemically gum arabic consists of a highly branched polysaccharide plus small amounts of protein matter.

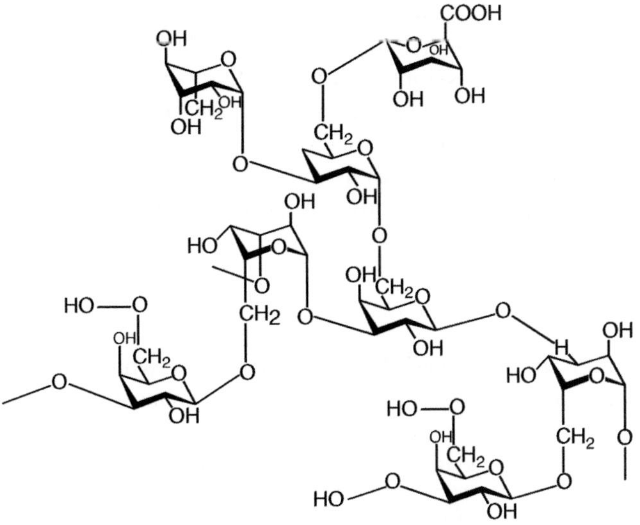

structure of gum arabic fragment

Gum arabic is highly valued as a thickener and emulsifying agent for oil–water systems, in many food and pharmaceutical products (Ali *et al.*, 2009). When present in the diet at levels of less than 10% the gum is completely digested and absorbed, it is without toxicity when ingested or when administered intravenously in humans (Miller, 1973). Traditionally twigs of gum arabic and related species have been used as dentifrices, a practice thought to assist dental hygiene and help prevent tooth decay. Studies have confirmed some antimicrobial effects, as well as enhancement of remineralization of tooth enamel (Ali *et al.*, 2009; Williams, 2011).

Acacia gums

Gum acacia is obtained from several species of Australian wattle (*Acacia* spp.) notably *Acacia decurrens*. Bark of this species was exported to the UK, and it was at one time included in the *British Pharmacopoeia*.

Wattle bark was used medicinally by indigenous Australians for coughs and colds, and to varying degrees for other ailments; it was also popular among white settlers as a domestic medicine mainly for gastrointestinal complaints. The bark is astringent due to its tannin content. Acacia bark was collected from trees 7 or more years old and then allowed to mature for a year before use, prepared either as an infusion or a decoction. The tea was mainly used for treatment of dysentery and diarrhoea (Williams, 2011).

The gum, which exudes from small injuries in the bark of the acacia, has also been used medicinally, although due to the higher tannin content the Australian product is regarded as inferior to that from Africa. The gum is collected by hand, sorted and dried. Purified samples are obtained by dissolving the crude gum in water, then recovering the polysaccharide by alcohol precipitation. Dissolved in water, the insipid gum makes a soothing syrup for inflamed mucous membranes.

Acacia gums are very soluble in water, yielding solutions of comparatively low viscosity in relation to those of most other polysaccharide gums. They are valued for their 'non-intrusive characteristics', emulsifying properties and stability, making them ideal vehicles for other ingredients in tablets, lozenges, antidiarrhoeal compounds and cough mixtures (Williams, 2011).

Seaweed gums

These are found in the 'leaves' (fronds) of seaweeds, which are conveniently classified according to colour – brown, red and green algae (Table 9.2).

Table 9.2. Common seaweeds and their polysaccharides

Seaweed genus	Common name	Polysaccharide type
Brown algae		
Laminaria	Kombu	Laminarin, alginates
Fucus	Bladderwrack	Fucans, fucoidans
Undaria	Wakame	Fucans, fucoidans
Ecklonia	Kelp	Fucoidans
Saccharina	Kombu	Fucoidans
Sargassum	Gulfweed	Fucoidan, alginates
Durvillaea	Bull kelp	Fucoidans, alginates
Red algae		
Chrondrus	Irish moss	Carageenan
Kappaphychus	Elkhorn sea moss	Carageenan
Eucheuma	Cottonii	Carageenan
Gelidium	Agarophytes	Agar
Gracilaria	Agarophytes	Agar
Porphyra	Nori	Porphyrans
Green algae		
Ulva	Sea lettuce	Ulvan – a disaccharide
Monostroma	Wittrock	Levoglucosan
Enteromorpha	Green seaweed	Ulvan – plus extra glucose

Brown algae are large seaweeds which inhabit the cold waters in both hemispheres. Their fronds contain alginates, non-sulfated polysaccharides valued for their gelling properties. Gelling strength is determined by the ratio of glucuronic acid (high gel strength) to mannuronic acid (low gel strength) (Iwaki *et al.*, 2012). **Alginates** are also used in pharmaceutical gastric antacids.

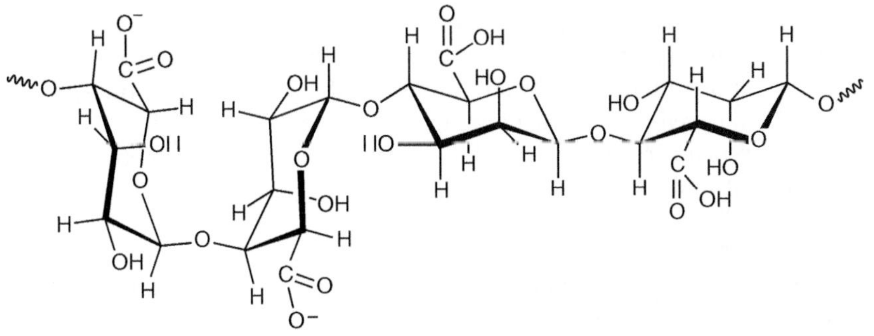

alginate – fragment of molecular structure showing
L-glucuronic acid (on left) and D-mannuronic acid (on right)

Fucoidans are non-gelling sulfated polysaccharides derived from brown macroalgae such as bladderwrack (*Fucus vesiculosus*, Fucaceae) and wakame (*Undaria pinnatifida*, Alariaceae). They are widely used as functional foods and dietary supplements for treatment of inflammatory disorders, such as osteroarthritis (Park *et al.*, 2010; Myers *et al.*, 2016). Antitumour effects have been demonstrated for fucoidans, based on inhibition of tumour cell proliferation, apoptosis stimulation, angiogenesis inhibition and immunostimulation. High sulfate content (from induced oversulfation) and low molecular size of the polysaccharides were found proportional to the antitumour effects, as well as for antiviral and radical scavenging activities (Qin, 2018; Shen *et al.*, 2018). In a clinical trial involving healthy adults, consumption of *U. pinnatifida* fucoidans demonstrated both immune stimulating and suppressing effects as indicated by altered immune cell counts (Ramberg *et al.*, 2010).

Red algae
Red algae inhabit warmer waters and they are the largest group of algae with over 6000 described species (Collén *et al.*, 2013). Much of the world production is from the Philippines and Indonesia. Polysaccharides are sulfated; they include **carrageenan**, agar and porphryans. Carageenan was originally derived from Irish moss (*Chrondrus crispus*, Gigartinaceae). It is now more commonly sourced from two Solieriaceae species: *Kappaphycus alvarezii* and *Eucheuma denticulatum*. Carrageenans have been classified as kappa (κ), iota (ι) and lambda (λ) groups depending on their level of sulfation.

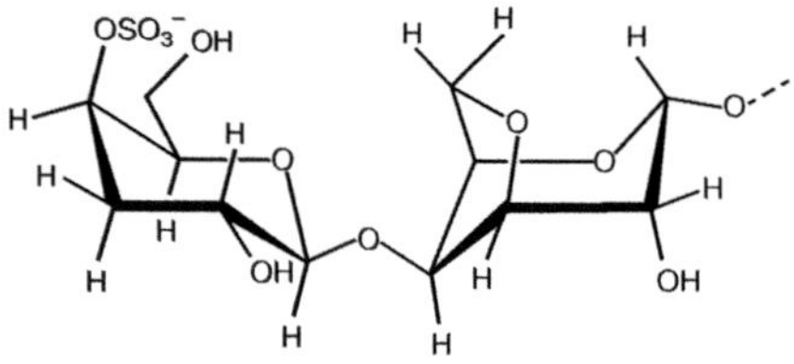

κ-carrageenan

Agar is derived mainly from species of *Gelidium* (Gelidiaceae) and *Gracilaria* (Gracilariaceae). The polysaccharide is treated with sodium hydroxide to reduce the sulfur content, thereby strengthening the gel strength of the agar.

Nori, the well-known sushi wrapping, is derived from the red algae, *Porphyra tenera* and related species (Bangiaceae). It is a very rich dietary source of vitamins A and B (Qin, 2018).

Green algae

Sea lettuces (*Enteromorpha* and *Ulva* spp., Ulvaceae) contain sulfated polysaccharides containing up to 8% protein, whose composition includes high levels of the amino acids asparagine and glutamine (Chattopadhyay *et al.*, 2007). Immunomodulating properties have been confirmed in studies with murine macrophages, the effects correlating with higher levels of sulfation of the polysaccharides, as previously noted for fucoidans (Leiro *et al.*, 2007).

Seaweeds are used as dietary ingredients and functional foods throughout the world. Their numerous benefits include prevention and treatment of obesity, diabetes, cancer, hyperlipidaemia and hypertension (Sun *et al.*, 2018).

Pectins

Pectins are complex galactouronic acid-based carbohydrates, the carboxylic acids are mostly esterified with methyoxyl or acetyl functional groups. Structurally pectins consist of a linear backbone of homogalacturonan, alternating with two types of highly branched rhamnogalacturonans (Lara-Espinoza *et al.*, 2018).

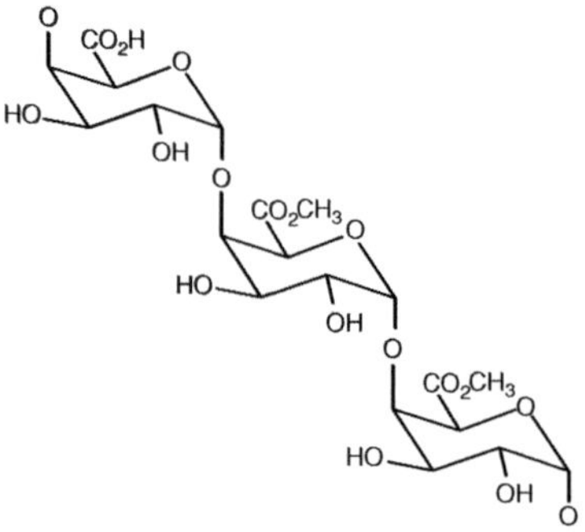

partial structure of esterified pectin

Pectins are found in the plant cell walls of most fruits and in other plant tissues. Commercially they are derived from citrus fruit, apple pomace and other sources. During fruit ripening an insoluble precursor is converted to soluble pectin and becomes gelatinous. It is used as setting agents for jams. Pectin has similar properties to gums – it adsorbs toxins, heavy metals, cholesterol and acts as a bulk laxative agent. Pectins are classified according to their degree of esterification; highly esterified (> 50% carboxyl groups) pectins have the best gelling properties.

Mucilages

Mucilages are long-chain polysaccharides that combine with water to form a slimy, semi-solid mass. They are not confined to specific plant parts, but may occur in:

- roots – comfrey (*Symphytum officinale*, Boraginaceae); marshmallow (*Althaea officinalis*, Malvaceae);
- leaves – coltsfoot (*Tussilago farfara*, Asteraceae);
- bark – slippery elm (*Ulmus rubra*, Ulmaceae); and
- seeds – psyllium (*Plantago ovata*, *Plantago afra*, Plantaginaceae).

Properties of mucilages

Mucilages have similar properties to those of seaweed gums, and the term is often applied to aqueous suspensions or solutions of gums and gelatinized starches. In physical terms they are hydrophilic (they attract water). In energetic terms they are cooling and sweet or bland.

Since they are, in the main, indigestible, their physical properties may be of more significance than their chemical properties. Even if some polysaccharides were broken down in the digestive tract, the breakdown products – sugars and uronic acids – have little pharmacological effect. They are, however, valued for certain pharmaceutical products where their gummy properties are helpful, such as in tablets and lozenges.

Local effects

The primary action is local, by direct contact with the surface of mucous membranes or skin. Here they produce a coating of slime that acts to soothe and protect exposed, irritated or inflamed surfaces of the oral and pharyngeal mucosa – a **demulcent** action. When this effect occurs on the skin it is referred to as an **emollient** action.

Mucilaginous herbs tend to be ideal for topical application for a wide variety of conditions ranging from bruises and swellings to irritable dermatological lesions. When applied topically, mucilaginous agents are soothing and emollient. Mucilages retain heat due to their hydrophilic properties, a characteristic often utilized in herbal preparations such as warm compresses, as they allow heat to penetrate the tissues progressively. Several herbs, including comfrey (*S. officinale*), slippery elm (*U. rubra*), marshmallow (*A. officinalis*) and linseed (*Linum usitatissimum*, Linaceae) are used in this way.

Gums and mucilages are invaluable aids in the management of irritable digestive disorders, especially where ulceration is a feature. Their relative indigestibility and hydrophilic properties create an influence on bowel behaviour, acting as:

- Laxative agents – the bulk effect results in increased peristalsis, drawing moisture into the colon producing a mild aperient effect. The best of these is psyllium seed or hulls from *Plantago* spp.
- Antidiarrhoeal agents – small quantities absorb excess water in the colon. Tannins if present have a binding effect (e.g. slippery elm powder).
- Demulcents (especially psyllium hulls) – are ideal for treatment of irritable bowel disorders. They are combined with antacid medications for relief of gastric reflux, oesophagitis and hiatal hernias. Slippery elm powder is particularly effective for these disorders.

Gums and mucilages also check fermentation and bacterial growth, and adsorb toxins and wastes helping their elimination from the body. Cholesterol is also lowered through this mechanism – a general property of water-soluble fibres such as those in oat bran. The hydrophilic effect produces a sensation of 'fullness' in the stomach without providing calories, hence the use of guar gum and others for appetite suppression. There is also a blood sugar lowering effect, observed in both diabetics and normal subjects. Psyllium use, for example, has been shown to reduce postprandial glucose concentrations and improve glycaemic and lipid control in individuals with type 2 diabetes (Kumar *et al.*, 2017).

The soothing, demulcent effects of gums and mucilages also benefit irritable states of the urinary and respiratory tract. Take, for example, the widespread use of marshmallow or Irish moss for bronchitis, and soothing preparations such as barley water for cystitis. Significant antitussive (cough suppressant) activity has been demonstrated *in vivo* for a range of mucilage-rich herbal medicines including *Aloe vera* and *Althaea officinalis*, and the antitussive effect was better or similar to that of non-narcotic pharmaceutical antitussives (Nosalova *et al.*, 2005). The main polysaccharide linked to the antitussive effect was rhamnogalacturonan.

While there is little difficulty in comprehending the strictly localized demulcent effect on the lining of the digestive tract, the mode of operation on organs with which the mucilage doesn't come into contact – respiratory tract, genitourinary tract and uterus – is believed to occur via vagal nerve reflex associations with the digestive tract, through an embryonic link in the nervous system (Mills, 1994).

Anti-inflammatory effects have also been demonstrated *in vivo* for aqueous (polysaccharide-rich) extracts of *A. officinalis* (Al-Snafi, 2013).

Mushroom polysaccharides

Polysaccharides are generally indigestible to humans, so their effects are not manifested in the bloodstream. However, their physical movement through the colon appears to trigger certain physiological reactions in humans, possibly due to an immune reaction via the Peyer's patches. Zymosan, a polysaccharide mixture isolated from Baker's yeast (*Saccharomyces cerevisiae*, Saccharomycetaceae), was found to stimulate macrophages and other immune cells in studies from the 1940s. Subsequently the active component of the zymosan mixture was isolated and characterized as a polysaccharide of the β-glucan structural type (Novak and Vetvocka, 2008).

Since then many polysaccharides, especially β-glucans, have been shown to have immunomodulatory effects and cytokine-stimulating activities. Cytokines can increase proliferation and differentiation of macrophages, T and B cells, and enhance cellular mechanisms of antitumour activity and antibody production. The polysaccharides tend to be large, insoluble, extensively branched molecules which are sometimes linked to a protein peptide (Turner, 1998). The following structural characteristics are closely associated with immunomodulating activities:

- high molecular weight;
- complex branching pattern;
- β-D-glucan type;
- glycosidal linkages: β–1,6′ β–1,3′ β–1,4′
- triple helix formation;
- presence of hydrophilic groups on outer surface of the helix;
- number of glucose residues;
- level of solubility; and
- low toxicity.

β-Glucan polysaccharides form helical structures – either single, double or triple helix. These are formed by three H-bonds in the C-2 position. The combination of high molecular weight, the stereochemistry of the sugars, diverse branching patterns and the presence of triple helix structures containing multiple receptor sites, allows the potential for a huge amount of biological information (Mahmoud, 2015). This in turn provides the flexibility required to interact with multiple types of regulatory immune cells, enabling the triggering of 'immune cascades'.

β-Glucans are also found in grains such as wheat, oats and barley; however, lower fibre grains are better adapted to modern food processing techniques, so that most commercial grain is now low in β-glucans. Hence mushrooms are regarded as the best source of these immune-stimulating compounds (Halpern, 2007) (see Table 9.3).

Medicinal mushrooms are biological response modifiers (BRMs) or adaptogens, and they activate immune responses in the host against malignant tumours (Wasser, 2002; Halpern, 2007; Mahmoud, 2015). General applications for β-glucan-rich mushrooms include as chemotherapy adjuvants in cancer patients, enhanced recovery of blood cell counts following radiology treatments, and reversal of myelosuppression (decreased bone marrow activity) following chemotherapy (Novak and Vetvocka, 2008). When mushroom polysaccharides are ingested as foods or in standardized capsules (guaranteed to carry specific polysaccharides), they are more likely to produce immunomodulating effects than when taken as tinctures or in other alcohol-based extracts. Hot water decoction is the most efficient method for breaking down the chitin-encrusted cell walls of mushrooms. Some of the more well-researched immunomodulating mushrooms are reviewed below.

Shiitake

Shiitake (*Lentinula edodes*, Omphalotaceae) is a wood-decaying mushroom from north-eastern Asia, where it is highly valued in terms of both flavour and nutritional status. The polysaccharide **lentinan** is a high molecular weight

Table 9.3. Selected medicinal mushrooms, their polysaccharides and therapeutic uses

Mushroom	Constituents (fruit body extract)	Therapeutic use
Agaricus blazei	β-Glucans 40%	Immune support, cancer
Ophiocordyceps sinensis (syn. *Cordyceps sinensis*)	β-Glucans 14%, cordycepic acid 6%	Adrenal tonic, fatigue, anti-ageing, neuroprotective
Trametes versicolor	Polysaccharopeptides, β-glucans 20%	Immune support, cancer, chemotherapy, HIV
Grifola frondosa	β-Glucans 20%	Immune support, HIV, dietary ingredient, diabetes, cancer
Ganoderma lingzhi syn. *G. lucidum*	β-Glucans 10%, heteroglycans, ganoderic acids 4%	Immune support, hepatitis, HIV, cardiac support, cancer
Lentinula edodes	β-Glucans 10% – lentinan, KS-2 – α-mannan-peptide	Immune support, antiviral, cholesterol management, dietary ingredient, cancer
Hericium erinaceus	β-Glucans 15%	Immune support, cognitive enhancement

(1,000,000 Da) β-(1→3) glucan, with side chains of β-(1→3) and β–(1→6) glucose residues, with no peptide. **KS-2**, a non-β-glucan polysaccharide with antitumour action and obtained from the mycelia only, contains mannose as the main component with a small amount of peptide (Bisen *et al.*, 2010). Lentinan is derived from both fruiting body and mycelia. It has pronounced antitumour activities in animals and is a potent interferon inducer. In clinical studies lentinan improved survival time in patients with inoperable gastric and recurrent breast cancers, given intravenously in conjunction with chemo-therapy. The mycelium extract was found to boost immunity in HIV patients, and to improve immune parameters and liver function in HIV patients (Bisen *et al.*, 2010). Potent immunomodulatory effects for lentinan were demon-strated in mouse models of inflammatory bowel disease and colitis-associated cancer (Liu *et al.*, 2018).

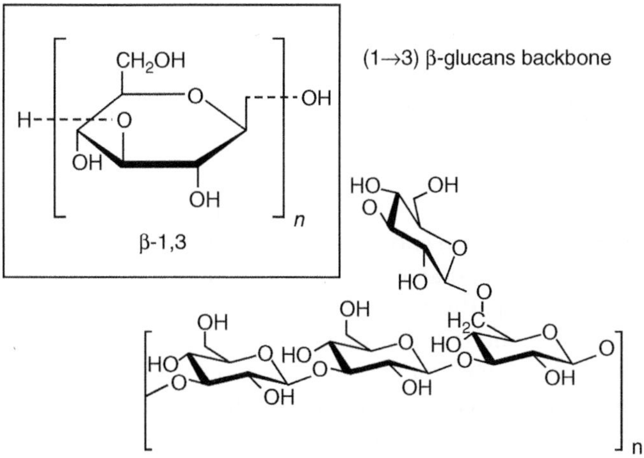

(1→3) β-glucans backbone

β-1,3

(1→3) β-glucans backbone with β(1→6) branching

lentinan (1→3)-glucose linkages in the backbone, (1→6) glucose side chains

Reishi

Reishi or lingzhi mushroom (*Ganoderma lingzhi* syn. *Ganoderma lucidum*, Ganodermataceae), are woody polypores characterized by having a double spore wall. Polysaccharides from reishi – especially the β-glucan fraction – have been found to have broad stimulatory effects on white blood cells, for example leukocytes, monocytes, macrophages and natural killer cells, which in turn lead to release of cytokines and lymphokines (interleukins, interferon, etc.). These are responsible for antitumour, antioxidant, anti-inflammatory, bacteri-cidal and immunomodulatory effects – useful in the treatment of autoimmune disorders, HIV-acquired immunodeficiency syndrome (AIDS) and cancer

(Ferreira *et al.*, 2015; Cör *et al.*, 2018). Apart from polysaccharides, *G. lingzhi* is also a rich source of mycochemicals including ganoderic acids (triterpenes), nucleotides, peptides, adenosine, organic germanium, ergosterol, resins, organic acids and lectins.

The antitumour effects of *G. lingzhi* are well established in terms of experimental research. In recent years several human studies have been conducted, these studies were subjected to a meta-analysis by the Cochrane Collaboration (Jin *et al.*, 2016). On the basis of five studies on Chinese subjects, the authors concluded there is no evidence to support the use of reishi alone in advanced cancer patients, but that better tumour response is observed when it was used as an adjuvant to chemo- and/or radiotherapy. The addition of reishi also diminished the immunosuppressive effects of chemo- and radiotherapy, enhanced T-lymphocyte counts, was well tolerated and led to improved quality of life among the subjects (Jin *et al.*, 2016).

Maitake

The so-called '**D-fraction**', a high-molecular-weight β-D-glucan from maitake mushroom (*Grifola frondosa*, Meripilaceae), consists of (1→3,1→6)-β-D-glucans, which account for 13% of the water-soluble polysaccharides in maitake (Nanba, 1993, 1997; He *et al.*, 2017). D-fraction has been extensively studied in animals and demonstrated enhanced macrophage, T cell and natural killer cell activity, increased interleukin(IL)-1 production and activation of the alternate complement pathway (Turner, 1998). Unlike lentinan in shiitake mushroom, the antitumour and immunomodulating properties of D-fraction are effective following both oral and intravenous administration (Nanba, 1997; Maximo and Andrea, 2020). Clinical studies show benefits to clients with different forms of cancer, when taken alone and more so when combined with chemotherapy (Nanba, 1997). As many as 47 different polysaccharides have been isolated from maitake fruiting body and mycelium, with considerable variation in the size and structures depending on the extraction methods used. For example, hot water extracts produced high levels of (1→4)-α-glucans (He *et al.*, 2017).

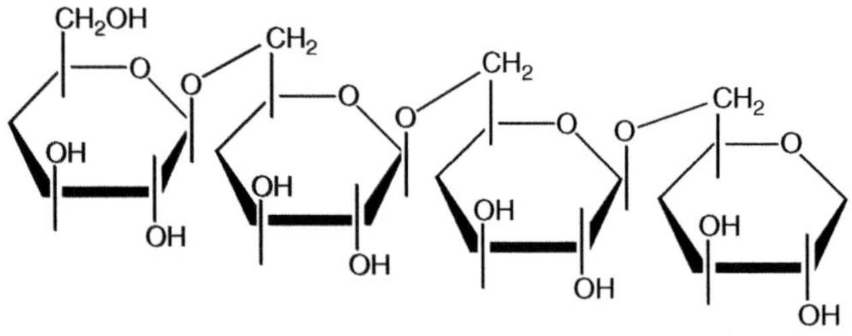

section of β-1,6'D-glucan

Coriolus

Trametes versicolor syn. *Coriolus versicolor* (Polyporaceae), known as coriolus or turkey-tail, is one of the most widely distributed polypore fungi in the world. Two significant polysaccharides are extracted from the mycelium of this species, polysaccharide-krestin (PSK) and polysaccharide-peptide (PSP), both of which are water soluble, and contain significant levels of protein and β-glucans. The polysaccharide complexes PSK and PSP have been used as adjunct treatments for a wide range of cancer types in Japan and China, respectively, the products having been approved as medicines for this purpose in both countries (Habtemariam, 2020). A meta-analysis of randomized controlled trials confirmed the efficacy of PSK in adjuvant immunotherapy for patients with curatively resected colorectal cancer (Sakamoto *et al.*, 2006). Apart from polysaccharides, *T. versicolor* also contains bioactive flavonoids, including several that are well known from medicinal plant research; these include baicalein, baicalin, quercetin, isorhamnetin, catechin and amentoflavone (Habtemariam, 2020).

Cordyceps

Cordyceps, an insect parasitic fungus from *Ophiocordyceps sinensis* (syn. *Cordyceps sinensis*, Cordycipitaceae) commonly known as caterpillar fungus, is found mainly in the Himalayan Mountains of Central Asia. Water-soluble polysaccharides comprise $(1{\rightarrow}3,1{\rightarrow}6)$-β-D-glucans, with potent immunostimulating action, demonstrated by elevated production of tumour necrosis factor (TNF-α) *in vitro*, and stimulation of the cytokines IL-4, IL-5 and IL-17 *in vivo* (Chen *et al.*, 2013). Immunostimulation is but one of cordycep's reputed therapeutic indications, in addition it has anti-ageing, restorative, antitumour, anti-inflammatory and neuroprotective actions (Chen *et al.*, 2013; Dhar *et al.*, 2015). These actions may be related to the many non-polysaccharide constituents present, including phytosterols, steroidal glycosides, amino acids and nucleotides such as adenosine. Cordycepin (3-deoxyadenosine), an adenosine analogue with antitumour (inhibits RNA/DNA synthesis), antiviral and antimalarial actions, is present only in low levels (Dhar *et al.*, 2015), however, the related fungus *Cordyceps militaris*, contains much higher levels of this active metabolite (Hawley *et al.*, 2020).

Plant polysaccharides

Immunomodulating polysaccharides are not confined to the fungal kingdom, they are also present in many plant-based medicines. Polysaccharides play significant roles in the activity of numerous herbs used in traditional Chinese medicine (TCM) and Japanese (Kampo) medicine, as Table 9.4 demonstrates.

Table 9.4. TCM and Kampo herbs containing bioactive
polysaccharides (based on a table in Chang, 2002)

Aloe spp.	*Bupleurum* root
Actractylodes spp.	*Glycyrrhiza* root
Cinnamomum bark	*Panax pseudoginseng*
Codonopsis pilosula	*Rehmannia* root
Curcuma zedoaria	*Salvia miltiorrhiza*
Isatis indigotica	*Zizyphus* fruit
Panax notoginseng	

There are also numerous examples linking polysaccharides from Chinese herbs with protective effects on the digestive system. Polysaccharides from *Panax ginseng* (Araliaceae), showed anti-ulcerogenic activity against induced ulcers (Sun, *et al.*, 1992) and **bupleuran 2II** from *Bupleurum falcatum* (Apiaceae) also exhibited significant anti-ulcer activity (Yamada *et al.*, 1991). Polysaccharides from dang gui (*Angelica sinensis*, Apiaceae) demonstrated protection of the gastrointestinal mucosa against ethanol and indomethacin (Cho *et al.*, 2000).

Water-soluble polysaccharides from *P. ginseng* showed non-specific immunomodulatory action in murine macrophages (Lim *et al.*, 2004) and antifatigue effects *in vivo* (Wang *et al.*, 2010). Furanose, a polysaccharide complex from *Panax quinquefolius* showed immunomodulatory effects when fed to healthy animals (Ramberg *et al.*, 2010). Polysaccharide-rich fractions of hot water extract of *Atractylodes lancea* (Asteraceae) stimulated proliferation of bone marrow cells mediated by Peyer's patch cells (Yu *et al.*, 1998). This study demonstrates the relationship of the gut-associated lymphoreticular tissue (GALT) and the intestinal immune system to the immunomodulating actions of polysaccharides (Hobbs, 2015).

The significance of polysaccharides with respect to the immune-enhancing properties of *Echinacea* spp. (Asteraceae) is controversial. From *Echinacea purpurea*, arabinogalactan and derivatives, and 4-O-methylglucuronoarabinoxylan have been shown to be immunostimulatory *in vitro*, and *in vivo* by intravenous administration (Glavaĉ *et al.*, 2012; Wink and van Wyk, 2017). There is almost no published research on *Echinacea angustifolia* polysaccharides.

Fructans

Fructans are water-soluble polymers of fructose stored in some plants as reserve material instead of starch. They have fewer sugar units and hence much lower molecular weight than starch. Unlike starch, fructans cannot be hydrolysed by digestive enzymes, rather they are selectively fermented by specific gut bacteria. In this way they function as prebiotics.

Studies indicate that fructans from foods such as rye, onions, garlic and artichoke tubers are digested in the large intestine by *Bifidobacterium* spp. and *Lactobacillus* spp. Health benefits include the production of short-chain fatty acids (acetate, propionate, butyrate, lactate), maintaining low pH, promoting healthy bowel flora and reducing bacterial pathogens such as *Biophilia* spp. (Cseke and Kaufman, 1999; Wilson and Whelan, 2017; Vandeputte *et al.*, 2017).

The branched fructans are found mainly in the grass (Poaceae) and lily (Liliaceae) families while linear fructans (specifically **inulin**) are particularly common in the Asteraceae (see Table 9.5). The tubers of dahlias and Jerusalem artichokes are sometimes used as a potato substitute. Fructans are composed almost entirely of fructosyl-fructose linkages, and in some cases glucose molecules are present in the chain (Cseke and Kaufman, 1999).

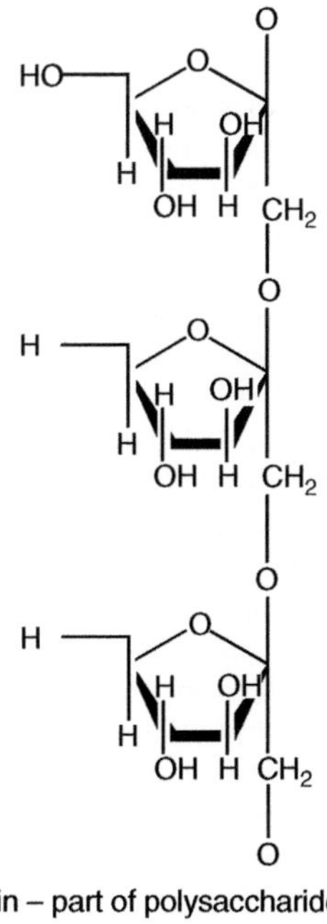

inulin – part of polysaccharide chain

Inulin contains up to 60 fructose residues linked by β-2,1 bonds with a terminal glucose unit. In human studies inulin influences key regulators of host metabolism and immunity, leading to healthier gut functioning (Wilson and Whelan, 2017; Zhang *et al.*, 2018). Inulin demonstrated antidiabetic effects in diabetic rats, linked to enhance serum GLP-1 levels and diminished IL-6 secretion, resulting in moderation of insulin tolerance (Zhang *et al.*, 2018).

Table 9.5. Inulin-containing species of herbs from the Asteraceae family

Botanical name	Common name
Arctium lappa	Burdock
Inula helenium	Elecampane
Cynara cardunculus var. *scolymus*	Globe artichoke
Cichorium intybus	Chicory
Echinacea spp.	Cone flower
Taraxacum officinale	Dandelion
Dahlia spp.	Dahlia
Helianthus tuberosus	Jerusalem artichoke

References

Ali, B.H., Ziada, A. and Blunden, G. (2009) Biological effects of gum arabic: a review of some recent research. *Food and Chemical Toxicology* 47(1), 1–8. doi: 10.1016/j. fct.2008.07.001

Al-Snafi, A.E. (2013) The pharmaceutical importance of *Althaea officinalis* and *Althaea rosea*: a review. *International Journal of PharmTech Research* 5(3), 1378–1385.

Barak, S. and Mudgil, D. (2014) Locust bean gum: processing, properties and food applications – a review. *International Journal of Biological Macromolecules* 66, 74–80. http://dx.doi.org/10.1016/j.ijbiomac.2014.02.017

Bisen, P.S., Baghel, R.K., Sanodiya, B.S., Thakur, G.S. and Prasad, G.B. (2010) *Lentinus edodes*: a macrofungus with pharmacological activities. *Current Medicinal Chemistry* 17(22), 2419–2430. https://doi.org/10.2174/092986710791698495

Bourbon, A.I., Pinheiro, A.C., Ribeiro, C., Maia, J.M., Teixeira, J.A. and Vicente, A.A. (2010) Characterization of galactomannans extracted from seeds of *Gleditsia triacanthos* and *Sophora japonica* through shear and extensional rheology: comparison with guar gum and locust bean gum. *Food Hydrocolloids* 24, 184–192.

Bruneton, J. (1995) *Pharmacognosy, Phytochemistry, Medicinal Plants.* Lavoisier Publications, Paris.

Chang, R. (2002) Bioactive polysaccharides from traditional Chinese medicine herbs as anticancer adjuvants. *Journal of Alternative and Complementary Therapies* 8, 559–565.

Chattopadhyay, K., Mandal, P., Lerouge, P., Driouich, A., Ghosal, P. and Ray, B. (2007) Sulphated polysaccharides from Indian samples of *Enteromorpha compressa* (Ulvales, Chlorophyta): isolation and structural features. *Food Chemistry* 104(3), 928–935. doi: 10.1016/j.foodchem.2006.12.048

Chen, P.X., Wang, S., Nie, S. and Marcone, M. (2013) Properties of *Cordyceps sinensis*: a review. *Journal of Functional Foods* 5(2), 550–569. https://doi.org/10.1016/j. jff.2013.01.034

Cho, C.H., Mei, Q.B., Shang, P., Lee, S.S., So, H.L., *et al.* (2000) Study of the gastrointestinal protective effects of polysaccharides from *Angelica sinensis* in rats. *Planta Medica* 66(4), 348–351. doi: 10.1055/s-2000-8552

Collén, J., Porcel, B., Carré, W., Ball, S.G., Chaparro, C., *et al.* (2013) Genome structure and metabolic features in the red seaweed *Chondrus crispus* shed light on evolution of the Archaeplastida. *Proceedings of the National Academy of Sciences of the United States of America* 110(13), 5247–5252.

Cör, D., Knez, Ž. and Knez Hrnčič, M. (2018) Antitumour, antimicrobial, antioxidant and antiacetylcholinesterase effect of *Ganoderma lucidum* terpenoids and polysaccharides: a review. *Molecules* 23, 649. doi: 10.3390/molecules23030649

Cseke, L.J. and Kaufman, P.B. (1999) How and why these compounds are synthesized in plants. In: Kaufman, P.B., Cseke, L.J., Warber, S., Duke, J.A. and Brielmann, H.L. (eds) *Natural Products from Plants.* CRC Press, Boca Raton, Florida, pp. 51–100. doi: 10.1201/9781420004472.ch2

Dhar, B.I., Srivastava, N., Himanshu, Jitindra, Sonika and Priyenka (2015) Chinese caterpillar mushroom *Cordyceps sinensis.* In: Janardhanan, K.K. and Ajith, T.A. (eds) *Developments in Medicinal Mushroom Biology and Therapeutic Properties.* Daya Publishing House, New Delhi.

Faria, S., Petkowicz, C., de Morais, S.A., Terrones, M.G., de Resende, M.M., *et al.* (2011) Characterization of xanthan gum produced from sugar cane broth. *Carbohydrate Polymers* 86, 469–476. doi: 10.1016/j.carbpol.2011.04.063

Ferreira, I.C., Heleno, S.A., Reis, F.S., Stojkovic, D., Queiroz, M.J., *et al.* (2015) Chemical features of *Ganoderma* polysaccharides with antioxidant, antitumor and antimicrobial activities. *Phytochemistry* 114, 38–55. https://doi.org/10.1016/j.phytochem.2014.10.011

Glavač, N.K., Iztok Jože Košir, I.J. and Kreft, S. (2012) Optimization and use of a spectrophotometric method for determining polysaccharides in *Echinacea purpurea. Central Europe Journal of Biology* 7(1), 126–131. doi: 10.2478/s11535-011-0091-z

Goswami, S. and Naik, S. (2014) Natural gums and its pharmaceutical application. *Journal of Scientific and Innovative Research* 3(1), 112–121.

Habtemariam, S. (2020) *Trametes versicolor* (synn. *Coriolus versicolor*) polysaccharides in cancer therapy: targets and efficacy. *Biomedicines* 8, 135. doi: 10.3390/biomedicines8050135

Halpern, G.M. (2007) *Healing Mushrooms. Effective Treatments for Today's Illnesses.* Square One, New York.

Hawley, S.A., Ross, F.A., Russell, F.M., Atrih, A., Lamont, D.J. and Hardie, D.G. (2020) Mechanism of activation of AMPK by cordycepin. *Cell Chemical Biology* 27, 214–222. https://doi.org/10.1016/j.chembiol.2020.01.004

He, X., Wang, X., Fang J., Yu, C. Ning, N., *et al.* (2017) Polysaccharides in *Grifola frondosa* mushroom and their health promoting properties: a review. *International Journal of Biological Macromolecules* 101, 910–921. doi: 10.1016/j.ijbiomac.2017.03.177

Hobbs, C. (2015) Mushroom medicine: challenges and potential. *Journal of the American Herbalists Guild* 12(2), 9–13.

Iwaki, Y.O., Hernandez Escalona, M., Briones, J.R. and Pawlicka, A. (2012) Sodium alginate-based ionic conducting membranes. *Molecular Crystals and Liquid Crystals* 554(1), 221–231. doi: 10.1080/15421406.2012.634329

Jin, X., Ruiz Beguerie, J., Sze, D.M. and Chan, G.C.F. (2016) *Ganoderma lucidum* (Reishi mushroom) for cancer treatment. *Cochrane Database of Systematic Reviews* (6): CD007731. https://doi.org/10.1002/14651858.CD007731.pub2

Kumar, D., Pandey, J., Kumar, P. and Raj, V. (2017) Psyllium mucilage and its use in pharmaceutical field: an overview. *Current Synthetic and Systems Biology* 5, 134. doi: 10.4172/2332-0737.1000134

Kyalangalilwa, B., Boatwright, J.S., Daru, B.H., Maurin, O. and van der Bank, M. (2013) Phylogenetic position and revised classification of *Acacia s.l.* (Fabaceae: Mimosoideae) in Africa, including new combinations in *Vachellia* and *Senegalia. Botanical Journal of the Linnean Society* 172(4), 500–523. https://doi.org/10.1111/boj.12047

Lara-Espinoza, C., Carvajal-Millán, E., Balandrán-Quintana, R., López-Franco, Y. and Rascón-Chu, A. (2018) Pectin and pectin-based composite materials: beyond food texture. *Molecules* 23(4), 942. https://doi.org/10.3390/molecules23040942

Leiro, J.M., Castro, R., Arranz, J.A. and Lamas, J. (2007) Immunomodulating activities of acidic sulphated polysaccharides obtained from the seaweed *Ulva rigida*. *International Immunopharmacology* 7, 879–888. doi: 10.1016/j.intimp.2007.02.007

Lim, T.S., Na, K., Choi, E.M., Chung, J.Y. and Hwang, J.K. (2004) Immunomodulating activities of polysaccharides isolated from *Panax ginseng*. *Journal of Medicinal Food* 7(1), 1–6. https://doi.org/10.1089/109662004322984626

Liu, Y., Zhao, J., Zhao, Y., Zong, S., Tian, Y., *et al.* (2018) Therapeutic effects of lentinan on inflammatory bowel disease and colitis-associated cancer. *Journal of Cellular and Molecular Medicine* 23, 750–760. doi: 10.1111/jcmm.13897

Mahmoud, A.E.E. (2015) *Bioactive Compounds in Edible Mushrooms*. Lambert Academic Publishing, Saarbrücken, Germany.

Maximo, A.B.D. and Andrea, B.G. (2020) Molecular mechanism induced by beta-glucans from maitake to recover T cell-subpopulations during immunosuppression. *IntechOpen* 10 June. doi: 10.5772/intechopen.90413. Available at: https://www.intechopen.com/online-first/molecular-mechanism-induced-by-beta-glucans-from-maitake-to-recover-t-cell-subpopulations-during-imm (accessed 6 December 2020).

Miller, L. (ed.) (1973) *Phytochemistry*, Vol. 1. Van Nostrand Reinhold Co., New York.

Mills, S. (1994) *The Essential Book of Herbal Medicine*. Viking Arkana, London.

Myers, S.P., Mulder, A.M., Baker, D.G., Robinson, S.R., Rolfe, M.I., *et al.* (2016) Effects of fucoidan from *Fucus vesiculosus* in reducing symptoms of osteoarthritis: a randomized placebo-controlled trial. *Biologics: Targets and Therapy* 10, 81–88.

Nanba, H. (1993) Antitumor activity of orally administered 'D-fraction' from maitake mushroom (*Grifola frondosa*). *Journal of Naturopathic Medicine* 4, 10–15.

Nanba, H. (1997) Maitake D-fraction: healing and preventing potentials for cancer. *Journal of Orthomolecular Medicine* 12(1), 43–49. Available at: http://orthomolecular.org/library/jom/1997/pdf/1997-v12n01-p043.pdf (accessed 6 December 2020).

Nosalova, G., Sutovska, M., Mokry, J., Kardosova, A., Capek, P. and Khan, M.T.H (2005) Efficacy of herbal substances according to cough reflex. *Minerva Biotecnologica* 17, 141–152.

Novak, M. and Vetvocka, V. (2008) β-Glucans, history, and the present: immunomodulatory aspects and mechanisms of action. *Journal of Immunotoxicology* 5, 47–57.

Park, S.B., Chun, K.R., Kim, J.K., Suk, K., Jung, Y.M. and Lee, W.H. (2010) The differential effect of high and low molecular weight fucoidans on the severity of collagen-induced arthritis in mice. *Phytotherapy Research* 24(9), 1384–1391. https://doi.org/10.1002/ptr.3140

Petri, D.F.S. (2015) Xanthan gum: a versatile biopolymer for biomedical and technological applications. *Journal of Applied Polymer Science* 132(23). https://doi.org/10.1002/app.42035

Qin, Y. (2018) Seaweed bioresources. In: Yin, Y. (ed.) *Bioactive Seaweeds for Food Applications*. Academic Press/Elsevier, Cambridge, Massachusetts, pp. 3–24. https://doi.org/10.1016/B978-0-12-813312-5.00001-7

Ramberg, J.E., Nelson, E.D. and Sinnott, R.A. (2010) Immunomodulatory dietary polysaccharides: a systematic review of the literature. *Nutrition Journal* 9: 54.

Sakamoto, J., Morita, S., Oba, K., Matsui, T., Kobayashi, M., *et al.* and Meta-Analysis Group of the Japanese Society for Cancer of the Colon Rectum (2006) Efficacy of adjuvant immunochemotherapy with polysaccharide K for patients with curatively resected colorectal cancer: a meta-analysis of centrally randomized controlled clinical trials. *Cancer Immunology, Immunotherapy* 55(4), 404–411. https://doi.org/10.1007/s00262-005-0054-1

Shen, P., Yin, Z., Qu, G. and Wang, C. (2018) Fucoidan and its health benefits. In: Yin, Y. (ed.) *Bioactive Seaweeds for Food Applications*. Academic Press/Elsevier, Cambridge, Massachusetts, pp. 223–238. https://doi.org/10.1016/B978-0-12-813312-5.00001-7

Sun, X., Matsumoto, T. and Yamada, H. (1992) Anti-ulcer activity and mode of action of the polysaccharide fraction from the leaves of *Panax ginseng*. *Planta Medica* 58(5), 432–435. doi: 10.1055/s-2006-961507

Sun, Z., Dai, Z., Zhang, W., Fan, S., Liu, H., *et al.* (2018) Antiobesity, antidiabetic, antioxidative, and antihyperlipidemic activities of bioactive seaweed substances. In: Yin, Y. (ed.) *Bioactive Seaweeds for Food Applications*. Academic Press/Elsevier, Cambridge, Massachusetts, 239–253. https://doi.org/10.1016/B978-0-12-813312-5.00012-1

Turner, D. (1998) The phytochemistry of immunity boosting polysaccharides. *British Journal of Phytotherapy* 5, 48–56.

Vandeputte, D., Falony, G., Vieira-Silva, S., Wang, M.S., Theis, S., *et al.* (2017) Prebiotic inulin-type fructans induce specific changes in the human gut microbiota. *Gut* 66(11), 1968–1974. doi: 10.1136/gutjnl-2016-313271

Wang, J., Li, S., Fan, Y., Chen, Y., Liu, D., *et al.* (2010) Anti-fatigue activity of the water-soluble polysaccharides isolated from *Panax ginseng* C. A. Meyer. *Journal of Ethnopharmacology* 130, 421–423. doi: 10.1016/j.jep.2010.05.027

Wasser, S.P. (2002) Medicinal mushrooms as a source of antitumor and immunostimulating polysaccharides. *Applied Microbiology and Biotechnology* 60, 258–274.

Williams, C. (2011) *Medicinal Plants in Australia*, Vol. 3. Rosenberg Publishing, Kenthurst, New South Wales, Australia.

Wilson, B. and Whelan, K. (2017) Prebiotic inulin-type fructans and galacto-oligosaccharides: definition, specificity, function, and application in gastrointestinal disorders. *Journal of Gastroenterology and Hepatology* 32(Suppl. 1), 64–68.

Wink, M. and van Wyk, B. (2017) *Medicinal Plants of the World*. CAB International, Wallingford, UK.

Yamada, H., Sun, X., Matsumoto, T., Ra, K., Hirano, M. and Kiyohara, H. (1991) Purification of anti-ulcer polysaccharides from the roots of *Bupleurum falcatum*. *Planta Medica* 57(6), 555–559. doi: 10.1055/s-2006-960205

Yu, K., Kiyohara, H., Matsumoto, T., Yang, H.C. and Yamada, H. (1998) Intestinal immune system modulating polysaccharides from rhizomes of *Actractylodes lancea*. *Planta Medica* 64(8), 714–719. doi: 10.1055/s-2006-957564

Zhang, Q., Yu, H., Xiao, X., Hu, L., Xin, F. and Yu, X. (2018) Inulin-type fructan improves diabetic phenotype and gut microbiota profiles in rats. *PeerJ* 6: e4446. https://doi.org/10.7717/peerj.4446

Alkaloids

10

Introduction

Alkaloids are alkaline, organic compounds containing a heterocyclic ring system with one or more nitrogen atoms, each connected to at least two carbon atoms within the ring. Most also contain oxygen. Unlike primary metabolites such as vitamins which occur in all plants, alkaloids have limited distribution, so that vitamins and hormones are excluded even though they may structurally comply with the definition. Most alkaloids are derived at least partly from various amino acids as their direct precursors, while some also have precursors derived from isoprene (terpenoids).

The isolation of morphine by Setürner in 1806 led to the discovery of several more alkaloids over the next 15 years, including emetine, piperine, caffeine and quinine. The term alkaloid was first applied by Meissner, a German pharmacist, and originally referred to all plant alkalis.

Found in roots, rhizomes, leaves, bark, fruit or seeds of 15–30% of all flowering plants, alkaloids are particularly common in certain families, such as the Fabaceae, Rubiaceae, Ranunculaceae, Apocynaceae, Solanaceae and Papaveraceae. Over 20,000 different alkaloids have been isolated from over 300 plant families (Mondal *et al.*, 2019). Over 150 alkaloids are known to occur in one species, *Catharanthus roseus* (Apocynaceae) (Dewick, 2009). The most widely occurring plant alkaloids are caffeine and berberine.

While higher plants are the major source of alkaloids, these chemicals are also known to occur in microorganisms, insects, the organs of mammals and in lower plants such as horsetail, algae and fungi (Kapoor, 1995).

Properties of alkaloids

Alkaloids are generally white or colourless crystalline solids. A few coloured alkaloids exist; these exceptions are sanguinarine (red), berberine (yellow) and chelidonine (yellow). With few exceptions (colchicine, berberine), alkaloids

  DOI: 10.1079/9781789243079.0010

are bases, turning red litmus paper blue and can be precipitated by various reagents, for example Mayer's reagent (mercuric chloride and potassium iodide). Some alkaloids give characteristic colour reactions with certain reagents. Most are susceptible to destruction by heat, some by exposure to air and light.

Soluble in organic solvents such as chloroform, ether and alcohol, most alkaloids are insoluble in water with a few exceptions that include ephedrine and colchicine. Alkaloidal salts are generally soluble in water and alcohol.

Nomenclature

All alkaloid names end in 'ine'. Otherwise the chemical names could have several possible origins:

- from the generic name of a plant – hydrastine (*Hydrastis canadensis*);
- from the specific name of a plant – cocaine (*Erythroxylon coca*);
- from the common name of a plant – ergotamine (ergot);
- from physiological activity – emetine (emetic action); and
- from a discoverer's name – lobeline (after L'Obel).

Where two or more similar alkaloids are present in a plant, prefixes, suffixes or other letters may be added to describe functional groups. For example, quinidine and hydroquinine are present along with quinine in *Cinchona* spp. (Rubiaceae).

Pharmacological actions

Plant alkaloids usually have profound physiological actions in humans with nervous system effects being the most prominent. Examples of some of the more dramatic actions of alkaloids are:

- analgesic/narcotic – morphine;
- mydriatic – atropine;
- miotic – pilocarpine;
- hypertensive – ephedrine;
- hypotensive – reserpine;
- bronchodilator – lobeline;
- stimulant – strychnine;
- antimicrobial – berberine; and
- antileukaemic – vinblastine.

Because they are so reactive, even at small closes, most alkaloid-rich plants are used sparingly in Western herbalism, if at all. Indeed, use of many alkaloid-containing plant species is restricted by law or listed on poison schedules. However, alkaloids – either as isolated compounds or their semi-synthetic derivatives – are widely used in pharmaceutical medicines. Much of our understanding of the mechanisms of neurotransmitters and receptor sites comes from research into the pharmacodynamics of alkaloids. The use of terms such as nicotinic and muscarinic receptors based on the alkaloids which bind to the receptors supports this idea.

Classification of alkaloids

Alkaloids are a large and diverse group of chemical compounds that defy easy classification. They are commonly grouped together according to their ring structures (Table 10.1). Two major divisions can be made:

1. Heterocyclic alkaloids – These are regarded as most typical.
2. Non-heterocyclic alkaloids – These are also known as protoalkaloids or biological amines (e.g. ephedrine, colchicine).

Table 10.1. Major alkaloidal groups, with examples and actions

Plant species	Pharmacological actions	Class	Example
Nicotiana tabacum	Adrenergic, CNS stimulant	Pyridine/piperidine	Nicotine
Lobelia inflata	Expectorant, bronchodilator	Pyridine/piperidine	Lobeline
Piper nigrum, Piper longum	Stimulant, hepatoprotective	Pyridine/piperidine	Piperine
Areca catechu	Vermicide, taenifuge	Pyridine/piperidine	Arecoline
Atropa belladonna	Anticholinergic, antisialagogue	Tropane	Hyoscyamine
Erythroxylon coca	CNS stimulant, anaesthetic, narcotic	Tropane	Cocaine
Datura metel, Datura stramonium	Anticholinergic, CNS depressant	Tropane	Scopalamine
Cinchona spp.	Antimalarial, antiarrhythmic, cardioactive	Quinoline	Quinine, quinidine
Berberis spp.	Antimicrobial, antiprotozoal, cholagogue	Isoquinoline	Berberine
Papaver somniferum	Sedative, analgesic, narcotic	Isoquinoline	Morphine
Chelidonium majus	Spasmolytic, cholagogue	Isoquinoline	Chelidonine
Peumus boldus	Spasmolytic, choleretic	Isoquinoline	Boldine
Carapichea ipecacuanha	Emetic	Isoquinoline	Emetine

Continued

Table 10.1. Continued

Plant species	Pharmacological actions	Class	Example
Sarothamnus scoparius	Oxytocic, cardiotonic, diuretic	Quinolizidine	Sparteine
Senecio jacobaea	Hepatotoxin	Pyrrolizidine	Sececionine
Symphytum spp.	Hepatotoxin	Pyrrolizidine	Symphytine
Rauwolfia serpentina	Sedative, antihypertensive	Indole	Reserpine
Claviceps purpurea	Vasoconstrictor, hypertensive	Indole	Ergotamine
Strychnos nux-vomica	CNS stimulant, deadly toxin	Indole	Strychnine
Aspidosperma quebracho	Aphrodisiac, stimulant	Indole	Yohimbine
Pilocarpus jaborandi	Miotic, cholinergic	Imidazole	Pilocarpine
Colchicum autumnale	Antimitotic, uric acid solvent	Alkaloidal amines	Colchicine
Ephedra sinica	Sympathetic stimulant	Alkaloidal amines	Ephedrine
Lophophora williamsii	Bronchodilator, hallucinogenic	Alkaloidal amines	Mescaline
Coffea arabica	CNS and sympathetic stimulant	Purine alkaloids	Caffeine
Thea sinensis	Bronchodilator, diuretic	Purine alkaloids	Theophylline
Paullinia cupana	CNS and sympathetic stimulant	Purine alkaloids	Guaranine
Solanum spp.	Steroid precursor, anti-inflammatory	Steroidal alkaloids	Solanine
Veratrum album	Cardiac depressant, antihypertensive	Steroidal alkaloids	Veratrine
Narcissus pseudonarcissus	Acetylcholinesterase inhibitor	Norbelladine	Galanthamine

CNS, central nervous system.

Pyridine-piperidine alkaloids

These alkaloids have their nitrogen atoms in typical six-membered carbon rings, which in the case of pyridine is a benzene ring. The precursors are generally **ornithine** and **nicotinic acid** although lobeline has a unique biosynthesis (see below).

Coniine, α-propylpiperidine, is found in spotted hemlock (*Conium maculatum*, Apiaceae), famous as the poison of Socrates. It is an enantiomeric piperidine structure with a short aliphatic side chain derived from the acetate/polyketide pathway, which imparts an oiliness to the compound. Coniine is very toxic and causes death by paralysis.

Nicotine, 1-methyl-2 (3-pyridyl) pyrrolidine, is derived from tobacco (*Nicotiana tabacum*) and other plants of the Solanaceae family. Because the precursor, nicotinic acid (vitamin B3), has widespread distribution in the plant kingdom, nicotine is also found in many remotely related plants, such as club mosses.

Pure nicotine is a colourless, oily alkaloid, while salts of nicotine are readily water soluble. The prime pharmacological alkaloid in tobacco is L-nicotine (0.5–10%), along with the secondary compounds **nornicotine**, **anabasine** and **nicotyrine**. Structurally, nicotine consists of a simple linking of pyridine and pyrollidine rings.

Tobacco is one of the most widely used entheogens – drugs that induce trance-like or divinatory states – throughout the American continent. The Australian pituri (*Duboisia hopwoodii*, Solanaceae), whose properties were exploited by the indigenous inhabitants of the central Australian desert, is a source of nicotine and nornicotine. Pituri can also be obtained from *Nicotiana* species native to Australia. In Westernized countries, tobacco has a very checkered history as a medicinal agent, having been initially promoted as a health tonic, then taken up for recreational use, and eventually demonized by the health authorities, while, currently, it is being used for the commercial production of a range of bio-pharmaceuticals (Sanchez-Ramos, 2020).

nicotine   acetylcholine – a neurotransmitter

From the pharmacological perspective, nicotine is cholinergic, with structural similarities to the neurotransmitter acetylcholine (Ach). So similar is the effect of nicotine to Ach on these cholinergic receptors they are known as nicotinic receptors. Different subtypes of nicotinic receptors have different physiological functions. The $\alpha4\beta2$ receptor subtype mediates the release of dopamine in the frontal cortex and mesolimbic area, known as the 'reward centre', responsible for the addictive properties of nicotine (Kennedy, 2014). Ach participates in the brain's cognitive functions, hence nicotine can improve attention, learning and memory while also acting as a neuroprotective agent (Posadas *et al.*, 2013).

Nicotine toxicity

Unfortunately, nicotine is extremely toxic, doses as low as 40 mg can kill an adult; however, the amount in cigarettes is very low, so that toxicity and addiction are associated with their long-term use. High doses of nicotine cause convulsions and death through respiratory failure, whereas chronic intake from smoking can lead to circulatory and cardiac disorders and death. Tobacco smoke ingredients such as tar and benzopyrenes are responsible for lung and other forms of cancer (Wink and van Wyk, 2008).

Nicotine and related alkaloids act as chemical defensive compounds against herbivores. The correlation between nicotine accumulation and its role with respect to herbivores has been studied in tobacco plants – a fourfold increase in the alkaloid content of leaves was observed in damaged versus non-damaged leaves (Roberts *et al.*, 2010).

Another nicotinic acid derivative is **trigonelline**, found in fenugreek seed (*Trigonella foecum-graecum*, Fabaceae) and unroasted coffee beans. Trigonelline has beneficial effects in diabetes and migraine headaches, it has antiviral and neuroprotective actions (Zhou *et al.*, 2012).

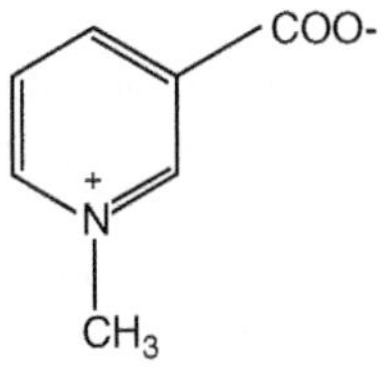

trigonelline – a pyridine alkaloid

Piperine (1-piperoylpiperidine) is found in black pepper (*Piper nigrum*, Piperaceae) and long pepper (*Piper longum*). Both peppers are of major importance in Ayurvedic medicine. Piperine has been shown to have hepatoprotective effects, though less potent than silymarin (Koul and Kapil, 1993). The amide side chain imparts high lipophilic (fat-soluble) properties, ensuring the compound is readily absorbed in the small intestine. Piperine has also been shown to enhance the bioavailability of other compounds, be they herbal or pharmaceutical. This interaction is thought to occur via an increased absorptive surface of the small intestine linked to alterations in membrane dynamics and permeation characteristics following ingestion of piperine (Khajuria *et al.*, 2002), as well as by the inhibition of drug metabolic enzymes and P-glycoprotein (Jin *et al.*, 2018).

The nuts of the betel palm (*Areca catechu*, Arecaceae) are widely used in many countries as a masticant. The nuts contain 15% tannins and **arecoline** (arecaidine methyl ester), an alkaloid which acts on the same receptors as nicotine.

Lobeline is found in *Lobelia inflata* (Campanulaceae) along with the related alkaloids **lobelanine** and **lobelanidine**. The main ring in lobeline is derived from lysine via piperidine, while the two benzene rings it contains derive from phenylalanine via the shikimic acid pathway (Samuelsson, 1992).

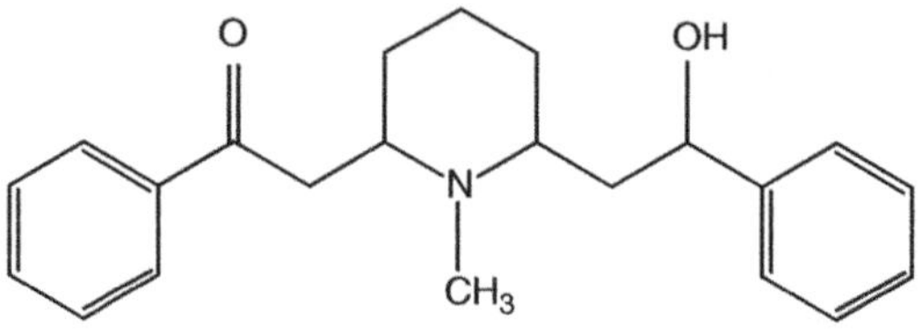

lobeline – a piperidine alkaloid

L. inflata is a relaxant and bronchodilator originally made famous by Samuel Thomson and the Physiomedical School of herbalists, however, in moderate high doses it becomes an emetic. Lobeline is a non-selective inhibitor of nicotinic, Ach and opioid receptors (Lee *et al.*, 2018) with limited use for treating alcohol, nicotine and methamphetamine dependence.

Quinoline alkaloids

Quinoline alkaloids are an example of a bicyclic ring system with the fusion of benzene and pyridine rings. Biosynthetically, they are related to indole alkaloids since both groups are derived from the same two precursors, tryptophan and loganin (a monoterpene iridoid) (Samuelsson, 1992), the difference being modification of the structure of the indole ring.

   Quinine (6-methoxycinchonine $C_{20}H_{24}N_2O_2$) is found in Peruvian bark (*Cinchona* spp., Rubiaceae), from trees that originate in the Andes mountains. The major species used are *Cinchona succirubra* and *Cinchona ledgeriana*. **Quinidine** is diastereoisomeric to quinine.

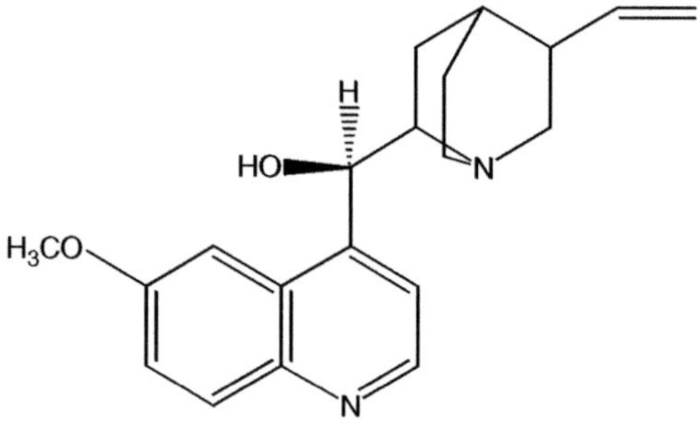

quinine

   Cinchona bark contains over 25 closely related alkaloids. The average yield is 6–7% (25% quinine, 5% quinidine), but higher yielding cultivars are commercially grown in plantations. The bark also contains cinchotannic acid (2–4%).

   Quinine is a complex alkaloid with several chiral centres. It is deadly to the protozoa *Plasmodium falciparum* the main causative agent in malaria, hence, it is used as the basis for antimalarial drugs.

Synthetic derivatives such as **chloroquinine**, which is more suitable as a prophylactic against malaria, gradually replaced quinine as the lead antimalarial drug (more recently with the unrelated terpenoid **absinthinin**). However, *P. falciparum* developed resistance to the synthetic analogues, leading to a reversion to quinine in some malarial-prone areas. Quinine use too has led to a degree of drug resistance. In one study the combination of quinine with quinidine and cinchonine was found to be up to ten times more effective *in vitro* against *P. falciparum* quinine-resistant and quinine-sensitive strains (Rasoanaivo *et al.*, 2011), suggesting there may still be a role for *Cinchona* extracts in malaria treatment.

Quinine is also used as a skeletal muscle relaxant (treatment of nocturnal cramps), and, traditionally, the bark extract has been used as a febrifuge for influenza and other acute infectious disorders. Quinine is very bitter. It is the main flavour component of tonic water.

Quinidine is a medical drug – indicated for cardiac complications, namely, arrhythmias, atrial flutter and fibrillation. It depresses myocardial excitability, conduction, velocity and contractility. It is used in the form of quinidine sulfate, though medical use has declined in recent years. Side effects of quinine drugs include tinnitus (cinchonism), skin rash, vertigo, unusual bleeding/bruising and visual disturbances.

Quinoline alkaloids are particularly common among plants of the Rutaceae family. Common rue (*Ruta graveolens*) contains 30 known alkaloids of the quinoline type, including, **arborinine** and **γ-fragarine** (Harborne and Baxter, 1993). Another Rutaceous plant genus with a broad alkaloidal spectrum is *Zanthoxylum*, of which there are over 200 species. Two North American species (*Zanthoxylum americanum* and *Zanthoxylum clava-herculis*) are the source of the herbal stimulant, prickly ash bark. *Z. americanum* contains the quinoline alkaloids **skimmianine** and **γ-fagrarine**, while both species contain isoquinoline alkaloids, including **nitidine** and **chelerythrine** (Nhiem *et al.*, 2020). Many more species grow in the Orient, and modern reviews of the genus identify hundreds of alkaloids from this genus and document therapeutic properties which include antimicrobial, anti-malarial, anti-inflammatory, antioxidant, hepatoprotective and, most notably, cytotoxic activities (Tian *et al.*, 2017; Nhiem *et al.*, 2020).

Camptothecin, from seeds, bark and leaves of the Chinese happy tree (*Camptotheca acuminata*, Nyssaceae), is a β-carboline derivative with a quinoline skeleton. Camptothecin can also be extracted from several other oriental herbs. It is as a broad-spectrum anticancer agent, acting by inhibition of DNA topoisomerase 1 (Zhao *et al.*, 2017; Shang *et al.*, 2018). Several semi-synthetic analogues of camptothecin are approved as cancer drugs.

campothecin – a pentacyclic quinoline alkaloid

Isoquinoline alkaloids

These alkaloids result from the condensation of a phenylethylamine derivative with a phenylacetaldehyde derivative. Both moieties are derived from the same precursors, either phenylalanine or tyrosine. The prototype alkaloid of this class is **papaverine**, found in several species of the poppy family (Papaveraceae).

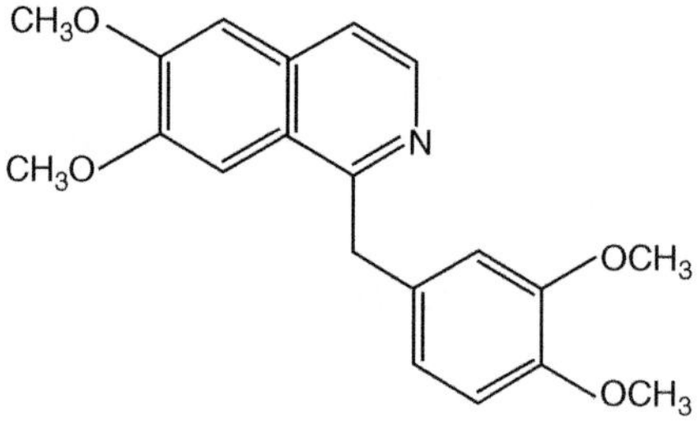

papaverine

Isoquinoline alkaloids are most frequently found in the Papaveraceae, Berberidaceae and Ranunculaceae families. This is a very large class of medicinally active alkaloids, which can be divided into the following subclassifications:

- morphinane alkaloids – morphine, codeine, thebaine (phenanthrene derivatives);
- benzylisoquinolines – papaverine, oxyacanthine, tubocurarine;
- protoberberines – berberine, hydrastine, palmatine;
- protopines – cryptopine, protopine;
- benzophenanthradine – chelidonine, sanguinarine;
- ipecac alkaloids – emetine, cephaeline; and
- aporphine alkaloids – boldine, magnoflorine, cordyline.

The properties of the various alkaloids in this class are extremely variable. Reported pharmacological actions include antispasmodic, antimicrobial, antitumour, antifungal, anti-inflammatory, cholagogue and hepatoprotective, antiviral, amoebicidal, antioxidant, cytotoxic and enzyme inhibitors.

Morphinane alkaloids – opiates

Opium is the dried latex obtained from incisions made in unripe fruits of the opium poppy (*Papaver somniferum*, Papaveraceae), a drug that has been used for many centuries, at least. The official opium drug is standardized to contain 10% morphine. Opium contains over 40 alkaloids, usually combined with meconic acid – a signature compound for identifying opium. Some major opium alkaloids (opiates) are listed below (Kapoor, 1995; Dewick, 2009):

- morphine (4–21%)–narcotic, analgesic and hypnotic;
- noscapine (4–8%)–antitussive, no narcosis;

- codeine (1–2.5%)–antitussive, mild narcotic and analgesic;
- papaverine (0.5–2.5%)–smooth muscle relaxant, antitussive; and
- thebaine (0.5–2%)–alternative source for synthesis of codeine.

The synthetic derivative **heroin (diacetylmorphine)** is more toxic and habit forming than morphine. It has a very short half-life.

Opioid receptors

The central-nervous-system effects of opiates occur through stimulation of opioid receptors in the G-protein coupled receptors (GPCRs) class. These receptors are widely distributed in animals, they respond to both endogenous transmitters (peptides) and ingested plant alkaloids.

The main opioid receptor types are:

- mu (μ) opioid receptor (MOR) – analgesic, euphoric;
- kappa (κ) opioid receptor (KOR) – sedative, dysphoric;
- delta (δ) opioid receptor (DOR) – anxiolytic; and
- nociceptin opioid peptide (NOP) receptor (also known as the nociceptin/ orphanin FQ (N/OFQ) receptor) – anti-analgesic.

The three classic receptor types (MOR, KOR, DOR) are further broken down into subtypes, each designated to a specific physiological action. While all subtypes provide a degree of analgesia, the MOR subtype μ_2, for example, also produces respiratory depression and causes itching (Dietis *et al.*, 2011). Endogenous opioid agonists include endorphins and enkephalins, peptides which are derived from tyrosine, hence there is similarity to the structure of morphinane alkaloids.

Opiate addiction is closely associated with the MOR receptors, which trigger euphoria via stimulation of the reward centres of the brain. Morphine and heroin are full MOR agonists. The most successful treatment for opiate addiction is methadone, which is also a full MOR agonist, but with a longer half-life than heroin it produces far milder withdrawal symptoms (Wang, 2019).

Morphine, $C_{17}H_{19}NO_3$, is a complex phenolic compound whose pentacyclic structure is derived from tyrosine. It is analgesic, narcotic and a powerful respiratory depressant previously used in cough elixirs. Morphine, still in use as the major drug for pain management in terminal illness is also known to induce euphoria and dependency in some people, anxiety and nausea in others. Other effects of morphine include a decrease in pupillary size, a reduction of hydro-chloric acid secretion in the stomach and constipation. Overdose of morphine can cause death through respiratory arrest (Kapoor, 1995; Dewick, 2009).

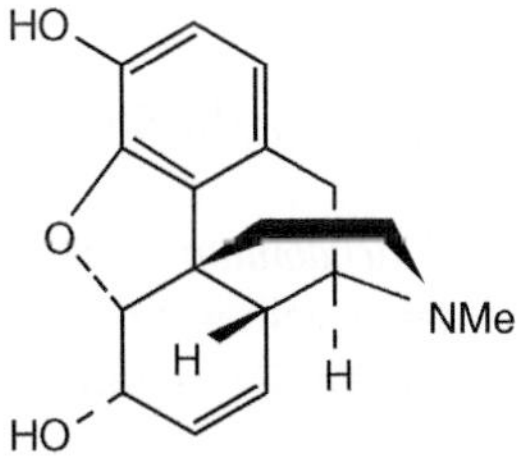

morphine – a pentacyclic morphinane alkaloid

Ipecac alkaloids

Ipecac is obtained from the dried rhizome and roots of *Carapichea ipecacuanha* (syn. *Cephaelis ipecacuanha*) (Rubiaceae), low shrubs indigenous to Brazil, Columbia and Nicaragua. Ipecac contains several isoquinoline alkaloids, including **emetine, cephaline** and **psychotrine**. Emetine hydrochloride is used as an antiprotozoal agent. Syrup of ipecac is an emetic and poison antidote.

Emetine is a gastrointestinal tract irritant, causing a reflex action leading to an increase in respiratory secretions. It acts as an emetic in higher doses and is often indicated for chronic bronchitis and whooping cough. Overdose can produce skin irritation, nausea, vomiting, bronchial oedema, diarrhoea, abdominal pain and rarely cardiac arrest and death (Wink and van Wyk, 2008). Ipecac and its alkaloids have potent antiviral effects, and in a recent study comparing the effects of a range of drugs being tested against coronoviruses, emetine was found to produce the highest level of inhibition (Bleasel and Peterson, 2020).

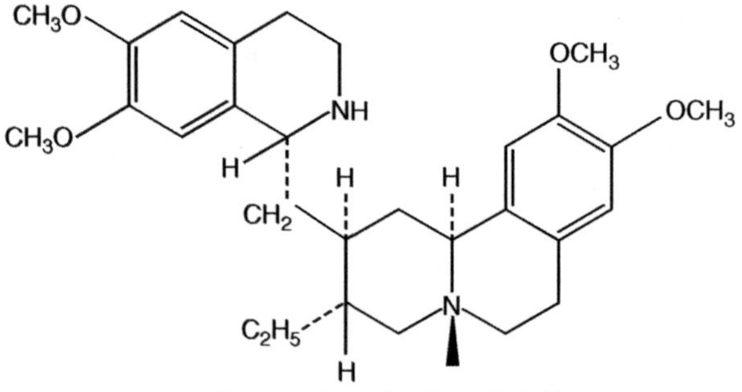

emetine – an isoquinoline alkaloid

Curare

Curare is a muscle relaxant, originally used as an arrow poison by Amazonian Indians. The traditional curare poison is made from various plant combinations that varied from tribe to tribe (Plotkin, 2020). Plant sources of curare include *Strychnos castelnaui* and other *Strychnos* spp. (Loganaceae) and *Chondodendron tomentosum* (Menispermaceae). **Tubocurarine**, a benzylisoquinoline dimer, is the major alkaloid in the curare plants. By competing with Ach at nicotinic receptor sites, it blocks nerve impulses at the neuromuscular junction and exhibits paralysing effects on skeletal muscles. Tubocurarine has been used as a muscle relaxant in surgical procedures; however, it has largely been replaced by synthetic analogues (Dewick, 2009).

Benzophenanthradines and protoberberines

The herb greater celandine *Chelidonium majus* (Papaveraceae) is a perennial European herb which exudes a yellow latex when any part of the plant is broken. Celandine contains over 40 alkaloids, including many from these two subcategories of isoquinoline type. **Chelidonine, berberine, sanguinarine, chelerythrine** and **protopine** are some examples (Zielińska *et al.*, 2018). Chelidonine is a yellow benzophenanthradine alkaloid. Pharmacological effects, mainly attributed to these alkaloids, include antimicrobial, antiprotozoal, choleretic, gastroprotective, hepatoprotective, antiproliferative, cytotoxic, analgesic, spasmolytic and anti-inflammatory (Gilca *et al.*, 2010; Zielińska *et al.*, 2018). Significant cytotoxic and proapoptotic activity for *C. majus* was demonstrated against various tumour cell lines, the activity being mainly attributed to the sanguinarine content (Och *et al.*, 2019).

In recent years concerns have been raised regarding possible hepatoxicity of *C. majus*, and while numerous instances of liver damage in humans have been attributed to it, in each case normal liver function was restored upon cessation of intake (Pantano *et al.*, 2017). Nevertheless, this herb should be taken only under professional supervision, and it is contraindicated to those with a history of liver disease.

Drug interaction considerations
Several alkaloids in this category contain a methylenedioxy moiety, a structural type that interacts with the haem portion of drug-metabolizing cytochrome P450 (CYP450) enzymes to form intermediate complexes, providing considerable risk for interactions with pharmaceutical drugs. They are mechanism-based inhibitors of CYP450 enzymes (Gurley *et al.*, 2012). Methylenedioxy moieties are not confined to isoquinolene alkaloids, but they are found in diverse compounds such as the lignan, **schisandrin, kavalactones** and the alkaloid, piperine.

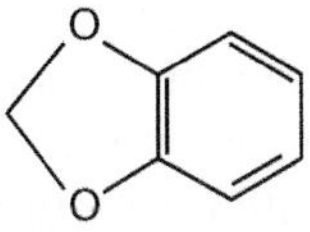

methylenedioxy moiety

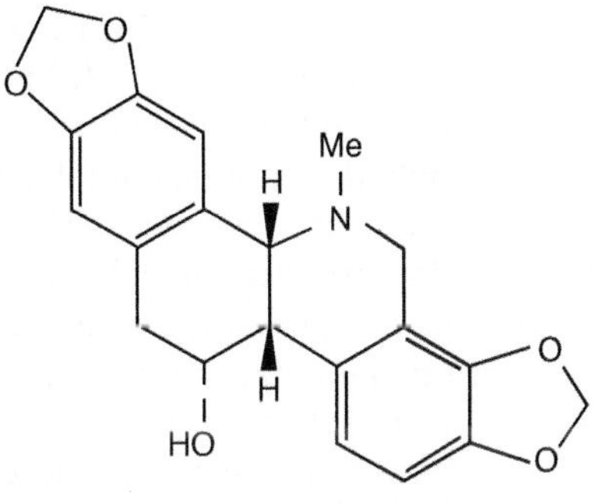

chelidonine – note the methylenedioxy moiety at each end

Berberine is a protoberberine quaternary alkaloid whose salts form yellow crystals. It is found, along with related alkaloids, in the following species: goldenseal (*Hydrastis canadensis*, Ranunculaceae), common barberry (*Berberis vulgaris*, Berberidaceae), Oregon grape (*Berberis aquifolium* syn. *Mahonia aquifolium*, Berberidaceae), Indian barberry (*Berberis aristata*, Berberidaceae) and golden thread (*Coptis chinensis*, Ranunculaceae).

berberine – a protoberberine isoquinoline alkaloid

Actions of berberine include: amoebicide, antibacterial, antifungal, cyto-toxic, hypotensive, anti-inflammatory, choleretic and antidiabetic (Wink and van Wyk, 2008; Tillhon *et al.*, 2012). Cytotoxic effects are less potent than for sanguinarine (Och *et al.*, 2019). Berberine has been shown to have a nega-tive inotropic effect on the heart, markedly reducing the atrial rate and has an antiarrhythmic action (Huang, 1993). Berberine and sanguinarine are potent inhibitors of DNA synthesis (repair and replication) – the likely mechanism for their antiviral effects. Add this to a variety of known effects, such as inhib-ition of protein biosynthesis and uncoupling of oxidative phosphorylation, berberine along with **palmatine** and sanguinarine can be regarded as potent allelochemicals that are toxic to bacteria and fungi, as well as other plants, insects and animals (Schmeller *et al.*, 1997).

Sanguinarine is a bright red alkaloid found mainly in bloodroot (*Sanguinaria canadensis*, Papaveraceae), a small herb from the forest understorey of North America. Both the alkaloid and the extracts of the herb have been used in proprietary dentifrices and mouthwashes, the most notable being Viadent® toothpaste and oral rinse products distributed by Colgate-Palmolive. Numerous studies have confirmed beneficial short- and long-term effects on plaque for-mation and gingivitis in periodontal patients taking sanguinarine-based dentifrices. However, such use has declined over the last two decades due to a possible association with leukoplakia, and it has been removed from the Viadent® product (Pengelly and Bennett, 2011). Experimentally sanguinarine is active against melanoma, multi-drug resistant human cervical cancer cell lines and drug-resistant malignant gliomas (Mondal *et al.*, 2019; Och *et al.*, 2019). It achieves this in part by intercalating DNA and inhibiting DNA topoisomerase 1 (Wink and van Wyk, 2008).

Boldine, (*S*)-2,9-dihydroxy-1,10-dimethoxyaporphine, is found in the leaves and bark of *Peumus boldus* (Monimiaceae), an evergreen tree native to Chile. Boldine imparts choleretic, cholagogue, antioxidant and smooth muscle relaxant properties to the herb (Speisky *et al.*, 1991). It is used primarily in the treatment of gallstones, especially for pain relief, digestive disorders and cystitis (Barnes *et al.*, 2007).

Tropane alkaloids

These are complex molecules containing pyrrolidine and benzene-ring structures. The pyrrolidine ring derives from either of the amino acids L-ornithine or L-arginine, via the polyamine intermediate putrescine. Extra carbons provided by acetyl-CoA lead to the formation of tropine. Tropine is esterfied with tropic acid, derived from the aromatic amino acid L-phenylalanine, providing the benzene ring required for the formation of **hyoscamine**, the base tropane alkaloid (Dewick, 2009; Roberts *et al.*, 2010).

The major tropane alkaloids are hyoscamine and **hyoscine** (syn. **scopolamine**). **Atropine**, the racemic form of hyoscamine, is found in *Atropa belladonna*, but in trace amounts only.

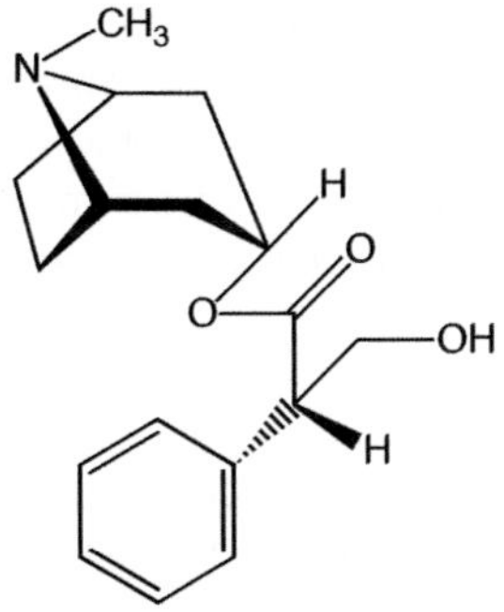

hyoscamine

These alkaloids and their near relatives are found in varying concentrations throughout the Solanaceae family, commonly known as 'nightshades'. Some of the main medicinal sources are:

- *Atropa belladonna* – deadly nightshade;
- *Datura stramonium* – thornapple;
- *Hyoscyamus niger* – henbane;
- *Mandragora officinarum* – mandrake; and
- *Duboisia myoporoides, Duboisia leichhardtii* – corkwood.

Nightshades have a long history of use as medicines, intoxicants and poisons in both the old and the new worlds. *Atropa, Hyoscyamus* and *Mandragora* are strongly associated with witchcraft in medieval Europe, *Datura* with shamanism in the Americas while *Duboisia* was made use of by Australia's Aborigines.

Pharmacological actions of hyoscamine and atropine

Hyoscyamine and atropine act on tissue cells innervated by post-ganglionic, cholinergic fibres of the parasympathetic nervous system; they are antimuscarinic and parasympathetic depressants. There are spasmolytic effects on bronchial and intestinal smooth muscles, caused by reducing the bronchoconstricting effects of Ach. The mydriatic effect (inhibits contraction of iris muscle) of hyoscamine is more pronounced than that of atropine. They both reduce salivary and sweat gland secretions and decrease intestinal tone thus reducing peristaltic contractions and producing antidiarrhoeal effects. Heart rate is increased due to suppression of vagal inhibition. Where hyoscyamine and atropine are central nervous system stimulants, scopolamine is a central nervous system depressant and is widely used for motion sickness transdermal patches (Wink and van Wyk, 2008).

Muscarinic antagonists are potentially useful in treating some disorders associated with dysfunction of the parasympathetic nervous system, notably epilepsy, Parkinson's disease and Alzheimer's disease (Kohnen-Johannsen and Kayser, 2019). Hyoscyamine can reduce rigidity and tremors in Parkinson's disease, a use first observed by the famous neurologist Jean-Martin Charcot in the 1870s, and which became the first official treatment for the disease. Hyoscyamine or belladonna were sometimes combined for this treatment with rye products containing ergot alkaloids, which are now known to be dopamine receptor agonists (Goetz, 2011).

Atropine has been used as an antidote in cases of poisoning by cholinesterase inhibitors such as the parasympathetic alkaloid, **phytostigmine**, and for organophosphates, which show up in insecticides and nerve gas used in biological warfare. Phytostigmine, in turn, is the antidote for poisoning by tropane alkaloids (Kohnen-Johannsen and Kayser, 2019).

Toxicity of tropane solanaceous alkaloids

Symptoms of poisoning by tropane-based solanaceous alkaloids include dilated pupils, impaired vision, dryness of skin, reduced secretions, extreme thirst, hot flushes and in higher doses, hallucinations, high fever, tachycardia and loss of consciousness. Atropine has an LD_{50} value (lethal dose required to kill 50% of a tested population of animals after a specified test duration) of 400 mg/kg, determined from studies on mice, while the lethal oral dose (LD_{100}) for adult humans is 10 mg (Wink and van Wyk, 2008).

Datura

The term datura includes *Datura stramonium*, also known as stramonium, jimson weed or thornapple and other species. Datura contains over 30 tropane alkaloids, mainly hyoscyamine and hyoscine, as well as their nitrogen oxides, **tigloylmeteloidine** and nicotine. Datura is occasionally used medicinally, where there are indications for asthma, pertussis, muscular spasm and for the excess salivation in Parkinson's disease (Bradley, 1992). The seeds and leaves may be smoked or burned for relief of asthma attacks. The bronchodilating effects of tropane alkaloids are utilized in modern prescription drugs such as Atrovent®.

Duboisia

Corkwoods (*Duboisia myoporoides* and *D. leichhardtii*) are trees/shrubs restricted to the east coast of Australia and adjacent areas. Corkwood leaves contain the highest levels of tropane alkaloids in the world, and since World War II have become the leading source of these alkaloids (Roddick, 1991; Kohnen-Johannsen and Kayser, 2019). *D. myoporoides* exists naturally in distinct chemical races, some of which have high levels of tropane alkaloids. In commercial cultivation a hybrid form of the species has been developed whose leaves contain up to 7% alkaloids (Evans, 1990; Roddick, 1991; Dewick, 2009). The most widely used *Duboisia* species by Aborigines, *D. hopwoodii* – also known as pituri – contains nornicotine and nicotine, and low levels of tropane alkaloids (Roddick, 1991; Kohnen-Johannsen and Kayser, 2019).

Cocaine

Cocaine is obtained from the leaves of the coca shrub, *Erythroxylon coca* (Erythroxylaceae), which is native to South America. Coca leaves have been chewed in Peru and Bolivia and beyond for at least 3000 years. Benefits ascribed to the practice include mild stimulation, pain management, alleviation of hunger and altitude sickness, as well as playing a central role in spiritual ceremonies across much of Latin America (Conzelman and White, 2016). There is little evidence of toxicity. However, due to the emergence of illicit cocaine production in order to meet the global demand for the drug, the use of coca leaf is now banned in most countries. Apart from cocaine, coca leaf contains numerous other alkaloids including **cinnamylcocaine, tropacocaine, benzoylecgonine** and **methylecgonine**, which are much less toxic than cocaine, along with carbohydrates, minerals and vitamins (Biondich and Joslin, 2016).

Upon local application, cocaine blocks nerve conduction, being useful as a local anaesthetic, particularly in dentistry. Cocaine analogues used for this purpose include **procaine** (**novacaine**) and **benzocaine**, however, their use has been largely superseded by cocaine-amide analogues such as **lidocaine** (Dewick, 2009). In large doses cocaine is a powerful cerebral stimulant and narcotic. Its adrenergic action is due to inhibition of the reuptake of noradrenaline, creating an amphetamine-like effect, though only of very short duration. Euphoric effects come from inhibition of the reuptake of dopamine and serotonin. Regular inhalation may cause destruction of mucous membranes such as those of the nose. Symptoms of overdose include dilated pupils, tachycardia, hypertension and death from cardiac arrest (Wink and van Wyk,

2008). Psychic dependence and tolerance may occur, since cocaine is quickly absorbed into the lungs, heart and brain with almost instant effects.

Quinolizidine alkaloids

These are also referred to as 'lupin' alkaloids since they were first discovered in *Lupinus* spp. They occur primarily but not exclusively in the Fabaceae family. The quinolizidine structure consists of two carbon rings with a shared nitrogen atom, formed from two molecules of the precursor amino acid, lysine.

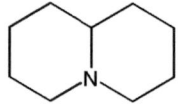

quinolizidine structure

The major quinolizidine alkaloids are **cytisine** and **lupinine** from lupins (*Lupinus* spp., Fabaceae), **myrine** from bilberry (*Vaccinium myrtillus*, Ericaceae) and **sparteine**, found in both Scotch broom (*Cytisus scoparius*, Fabaceae) and greater celandine (*Chelidonium majus*, Papaveraceae).

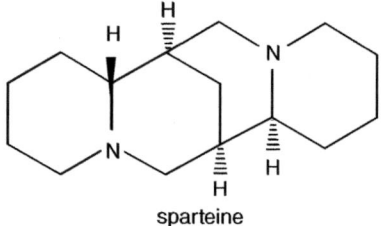

sparteine

Sparteine is a tetracyclic, oxygen-free alkaloid, derived from three lysine molecules. It is oxytocic, a cardiac stimulant and diuretic and binds strongly to muscarinic receptors and nicotinic receptors. Like nicotine, it is a peripheral vasoconstrictor and hypertensive and has been used as an aid to quit smoking (Wink and van Wyk, 2008). Sparteine has a similar cardiac action to quinidine and it has been used for cardiac arrhythmia. The action of broom in this regard is modified by the presence of flavonoids. It has an ergot-like action on the uterus and has been used as a substitute for ergot drugs. However, overdose of sparteine can be fatal as very high doses may cause strychnine-like convulsions and death from suffocation (Wink and van Wyk, 2008). Sparteine-containing herbs are contraindicated during pregnancy.

Pyrrolizidine alkaloids

Pyrrolizidine alkaloids (PAs) contain two fused five-carbon rings, known as necines, in which a nitrogen atom is common to both rings, as well as one or

more necic acid functional groups. The precursors are two ornithine molecules, which the pyrrolizidine structure incorporates via different intermediates
(Dewick, 2009). A hydrogen atom usually occurs opposite the nitrogen atom
in the α-position, with a hydroxymethyl substitute at the adjacent C-l position.
Most pyrrolizidines have diester groups at C-l and C-7 – these may be non-cyclic,
for example **echimidine**, or macrocyclic, for example **senecionine**. Few therapeutic effects have been postulated for PAs, the research interest is mainly
confined to their toxicity.

senecionine – a pyrrolizidine
macrocyclic diester

PA toxicity is mainly associated with the unsaturated macrocyclic diester
type, also referred to as 1,2-dihydropyrrolizidine alkaloids (DHPA). They are
subject to liver biotransformation in humans through oxidation by CYP450
enzymes and the production of toxic pyrrole esters, which can bind to proteins
and DNA forming adducts (covalent modifications of DNA) (Moreira *et al.*,
2018). These toxic metabolites are responsible for venous occlusive disease,
also known as hepatic sinusoidal obstruction syndrome (HSOS) – a form of
hepatotoxicity that can be fatal. While acute effects from PA ingestion may
occur, the effects of chronic exposure are the most well documented, characterized by necrosis, fibrosis, cirrhosis and proliferation of the bile duct epithelium
(Moreira *et al.*, 2018). The cyclic diester forms are considered the most toxic,
the non-cyclic diesters are of intermediate toxicity, and the monoesters are the
least toxic (Roeder *et al.*, 2015; Chen *et al.*, 2017; Moreira *et al.*, 2018). PA alkaloids
also occur as water-soluble nitrogen oxides, a non-toxic form.

Herb teas are susceptible to contamination by PA-containing weeds; this has
led to increased screening of herbal drugs including herb teas for the presence
of PAs (Chen *et al.*, 2017). An additional concern is PA contamination of grains,
honey, pollen and milk. There are said to be over 600 PA-containing plant species,
mainly in the Boraginaceae, Papilionaceae and Asteraceae families.

Medicinal plants with discernable pyrrolizidine levels are scheduled or
restricted in many countries (EFSA Panel on Contaminants in the Food Chain
(CONTAM), 2011; Moreira *et al.*, 2018). The plants include comfrey (*Symphytum*
spp.), borage (*Borago officinalis*) and lungwort (*Pulmonaria officinalis*) in the
Boraginaceae, coltsfoot (*Tussilago farfara*), butterbur (*Petasites* spp.) and liferoot
(*Packera aurea* formerly *Senecio aureus*) in the Asteraceae. Although diesters are

present in comfrey, the main PAs present are retronecine monoesters, the least toxic category (Rode, 2002). Recently low levels of monesters **lycopsamine** and **intermedine** were isolated from the American herb, boneset (*Eupatorium perfoliatum*, Asteraceae) (Upton and Petrone, 2019). While the risk posed by this herb is very low, only short-term use is recommended.

Indole alkaloids

Indoles are a very large group of alkaloids whose basic structure contains a pyrrole ring fused to a benzene ring. Biosynthesis of pure indoles involves the amino acid **L-tryptophan** as precursor, while that of a major subgroup of the indoles, including the *Catharanthus* and *Rauwolfia* alkaloids, involves a second precursor – the monoterpene iridoid **loganin**. L-tryptophan is also the precursor to alkaloid-like neurotransmitters including **tryptamine, serotonin** (5-hydroxytryptamine) and **melatonin**.

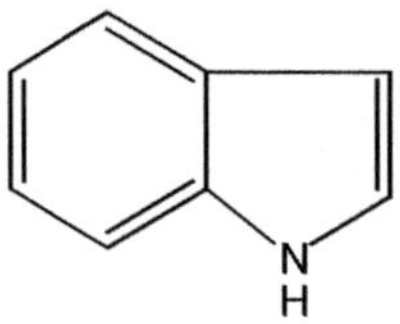

indole 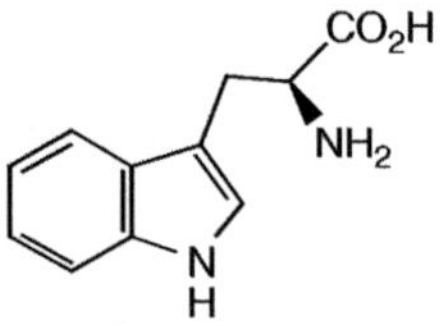   tryptamine – a trace amine

The indole alkaloid structures typically involve multiple ring systems, often complex in character. They form the basis of several pharmaceutical drugs and some of the most potent hallucinogens as well as poisonous compounds such as strychnine.

Of the indole alkaloids used in medicine and pharmacy, the majority are found in members of the family Apocynaceae (e.g. *Rauwolfia, Vinca, Catharanthus* and *Alstonia* spp.). Other families in which they occur are the Loganaceae, Rubiaceae and Convovulaceae. Indole alkaloids are represented in the fungal kingdom by the ergot alkaloids (*Claviceps* spp., Clavicipitaceae), the *Psilocybe* genus (Hymenogastraceae), *Panaeolus* spp. and *Copelandia* spp. (Bolbitaceae).

Reserpine is a tryptamine-derived alkaloid found in the roots of *Rauwolfia serpentina* (Apocynaceae) along with the related alkaloids **resicinnamine, deserpidine** and **ajmaline**. The main actions are hypotensive, sedative and tranquillizing. Ajmaline is also of benefit for heart arrhythmias (Dewick, 2009).

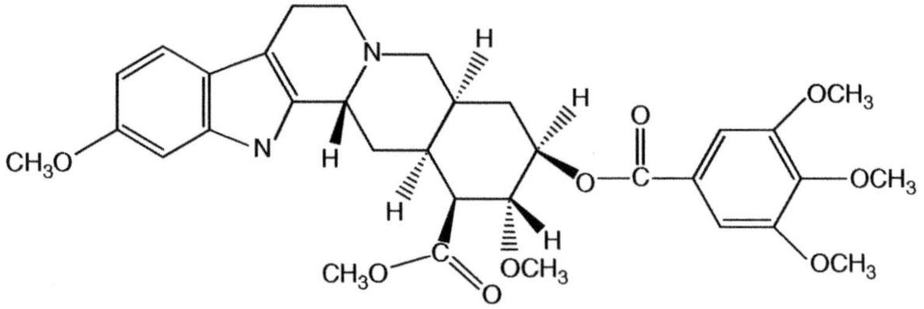

reserpine

The primary actions of reserpine alkaloids are caused by inhibition of nor-adrenaline and depletion of amines in the central nervous system. While the hypotensive effects have a slow onset, their duration is long, and the effective dose is sufficiently low to limit any side effects. It was previously used as a tranquillizer, however, the much higher doses required to obtain this effect often resulted in depression and Parkinsonism-like symptoms (Huang, 1993). As a result, *Rauwolfia* has become a restricted herb and is rarely used in current herbal prescribing.

Alstonia constricta (Apocynaceae), bitter bark, is one of the best of the bitter tonics and febrifuge remedies to come from Australia. However, it is a uterine stimulant and should be avoided during pregnancy. It contains several antihypertensive alkaloids of the indole class – **alstonine, alstonidine** and small amounts of reserpine. Widely distributed in Southern Asia and Northern Australia, *Alstonia scholaris* contains a host of indole alkaloids, including **19-epischolaricine, vallesamine, picrinine** and **alstoscholarine**. Indole alkaloids from this species have demonstrated antitussive, antiasthmatic, anti-inflammatory, analgesic and expectorant activities (Pan *et al.*, 2016).

Periwinkle

As noted previously, well over 150 alkaloids have been isolated from the rose or Madagascar periwinkle (*Catharanthus roseus*, Apocynaceae), popular garden plants. However, the alkaloids only occur naturally in very small quantities. **Vinblastine** and **vincristine** are antineoplastic, used as chemotherapy treatment in childhood leukaemia, Hodgkin's and non-Hodgkin's lymphomas, also breast, lung and testicular cancers. They inhibit mitosis of cancer cells by disrupting microtubule function (Mondal *et al.*, 2019). Like other chemotherapeutic agents, these alkaloids are toxic, adverse effects include nausea, vomiting, constipation, neutropenia, dyspnea and chest pains (Anitha, 2016).

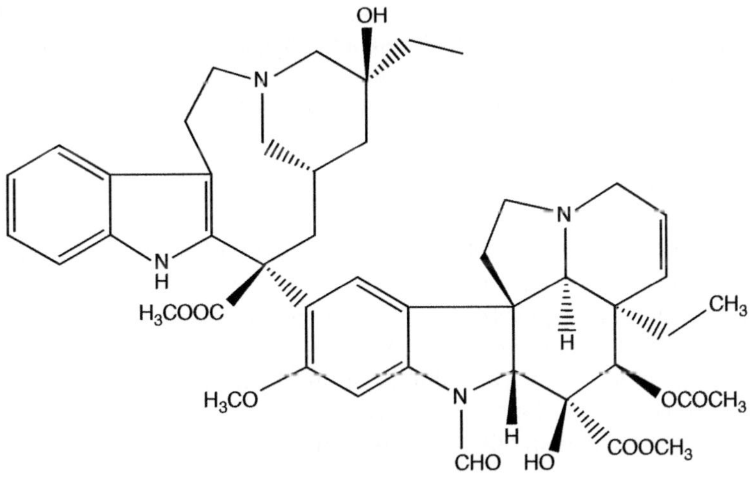

vinblastine – a dimeric indole alkaloid

The common periwinkles (*Vinca minor* and *Vinca major*, Apocynaceae) contain numerous indole alkaloids including **vincamine**, **majdine** and **majoridine**. Their actions are antihaemorrhagic and astringent. *Vinca* spp. do not contain the antineoplastic alkaloids found in *Cantharanthus* (Wren, 1988).

Kratom and iboga

Kratom (*Mitragyna speciosa*, Rubiaceae) is an Asian herb that contains over 50 indole alkaloids in its leaves, most notably **mitragynine**, a tetracyclic indole alkaloid with a methoxy group, plus a related class of alkaloids, pentacyclic oxindoles (Flores-Bocanegra *et al.*, 2020). Kratom has been widely used for treatment of opioid and alcohol dependence, particularly in Thailand and Malaysia. Mitragynine has a high affinity for MORs (mu opioid receptors), and it also binds to KORs and DORs (kappa and delta opioid receptors, respectively). *In vivo* studies indicate antinociceptive, antidepressant and narcotic effects, which may assist in treatment of opioid dependence, however, kratom also has potential to become addictive (Suhaimi *et al.*, 2016). In a recent pharmacokinetics and safety study with dogs, oral intake of mitragynine produced mild sedation and anxiety, but was otherwise well tolerated (Maxwell *et al.*, 2020).

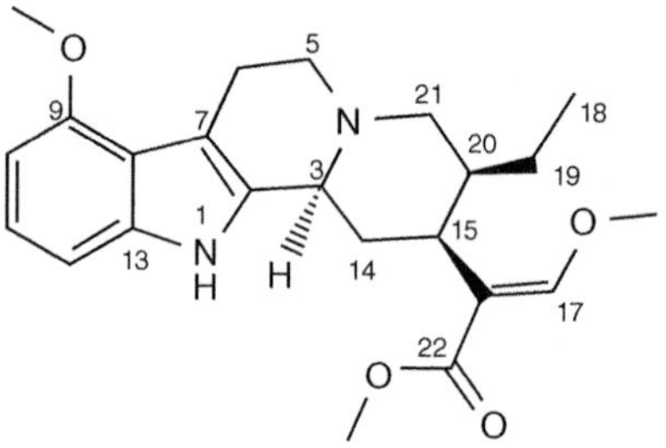

mitragynine – a tetracyclic indole alkaloid

Another alkaloidal plant with potential for treating drug addiction is iboga (*Tabernanthe iboga*, Apocynaceae), an African shrub whose roots contain the monoterpenoid indole alkaloids **ibogaine**, **ibogamine** and **tabernanthine**. Iboga is a psychoactive hallucinogen, which in high doses can cause convulsions, paralysis and death (Wink and van Wyk, 2008).

Ergot alkaloids

Ergot (*Claviceps purpurea*, Clavicipitaceae) is a fungus with a sclerotium (fruiting body) that produces over 20 indole alkaloids, derivatives of lysergic acid rather than tryptamine. Ergot grows on cereal plants (rye, wheat, etc.) and wild grasses around the world. Ergotism is a toxic response to ingesting ergot-contaminated grain, and manifests either as painful spasms of the limb muscles leading to epileptic-like convulsions (St Anthony's fire), or as vomiting and diarrhoea leading to gangrene of the toes and fingers. Both syndromes can lead to fatalities.

Ergotamine constricts peripheral blood vessels and raises blood pressure, by binding to α-adrenergic and serotonergic receptors. It is used in treatment of migraine headaches, often in combination with caffeine. **Ergonavine** is oxytocic and vasoconstrictive and is used as treatment or a preventative for post-partum haemorrhage. **Bromocriptine**, a semi-synthetic derivative of lysergic acid, stimulates dopaminergic receptors, and is used for treatment of Parkinson's disease (Dewick, 2009).

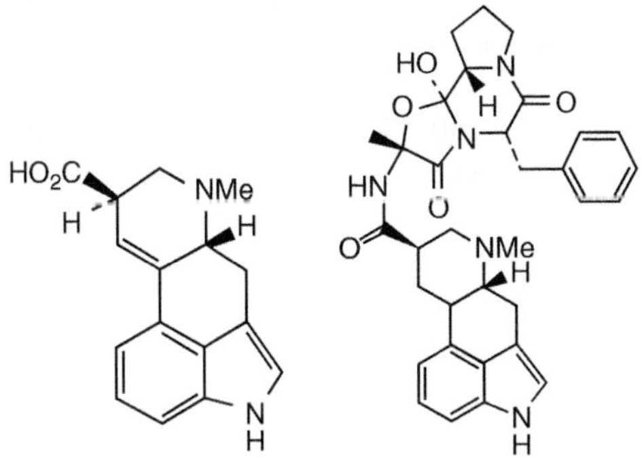

lysergic acid   ergotamine

Hallucinogens and entheogens

LSD (**lysergic acid diethylamide**), a hallucinogen, is another semi-synthetic lysergic acid derivative, whose actions are partially explained by agonistic effects on serotonergic, dopaminergic and adrenergic receptors (Kennedy, 2014). **Lysergic acid amide** is a natural derivative found in the seeds of some members of the morning glory (Convovulaceae) family such as *Ipomoea violacea*. This and other species, known collectively as ololiuqui, are used in Mexico as entheogens (Kennedy, 2014). As hallucinogens, they are very mild by comparison to LSD and their potential for toxicity restricts any recreational value (Dewick, 2009).

The *Psilocybe* mushroom alkaloids **psilocybin** and **psilocin**, structurally very similar to their precursor tryptamine, are further examples of entheogens but with definite hallucinogenic properties. These alkaloids readily cross the blood–brain barrier, and they are strong agonists to 5-HT$_{2A}$ receptors. Psilocybin is currently under clinical investigation for treatment-resistant depression (Carhart-Harris *et al.*, 2017; Roseman *et al.*, 2018) and obsessive compulsive disorder (OCD) (Jacobs, 2020).

Possibly the most powerful natural hallucinogen comes from ayahuasca, a traditional Amazonian mixture of plant extracts whose main ingredients are the liana (jungle vine) *Banisteriopsis caapi* (Malpighiaceae) and the shrubby *Psychotria viridis* or related species (Rubiaceae). *Psychotria* contains the hallucinogenic indole alkaloid **N,N-dimethyltryptamine** (**DMT**) and dervatives such as **5-MeO-DMT**, while *B. caapi* contains β-carboline indole alkaloids, mainly **harmine** and **harmaline**, which are serotonin antagonists (Wink and van Wyk, 2008; Kennedy, 2014; Plotkin, 2020).

Strychnine comes from the seeds of *Strychnos nux-vomica* (Loganiaceae) known as nux vomica or the poison nut tree. It is a powerful stimulant to the central nervous system in low doses, but becomes deadly poisonous in larger doses (> 90 mg in adults), causing an exaggeration of reflexes and tonic convulsions, and death by asphyxiation or cardiac arrest (Wink and van Wyk, 2008). **Brucine**, a related but slightly less toxic alkaloid in *Strychnos* spp., is famous as the source of poison in the classic novel, *The Count of Monte Cristo* by Alexander Dumas. Interestingly, the antidote for both strychnine and brucine poisoning is **curare**, a paralyzing agent extracted from species of *Strychnos*.

Yohimbine, another tryptamine derivative, comes from the barks of *Aspidosperma quebracho* (Apocynaceae) and *Pausinystalia johimbe* (Rubiaceae). These plants are reputedly aphrodisiacs. Yohimbine blocks α-adrenergic, serotonergic and dopaminergic receptors, and is a vasodilator (Dewick 2009; Kennedy, 2014). Yohimbine and strychnine have been combined in pharmaceutical preparations for use as aphrodisiacs and nerve tonics.

Steroidal alkaloids

Steroidal alkaloids are derived from triterpenoids, the precursor being cholesterol. They are distinguished from other compounds in that class by the presence of a nitrogen atom, derived from L-argenine. Most steroidal alkaloids have C_{27} structures, which occur as glycosides (also known as glycoalkaloids), having similar actions to triterpenoid saponins.

The largest subclass is the **cholestanes**, found only in the Solanaceae family. In unripe potatoes (*Solanum tuberosum*, Solanaceae) the common glycoalkaloids **α-solanine** and **α-chaconine** are derived from the aglycone **solanidine**. Glycoalkaloids are mainly concentrated in unripe fruits (including tomatoes and green potatoes), but disappear in the ripening process.

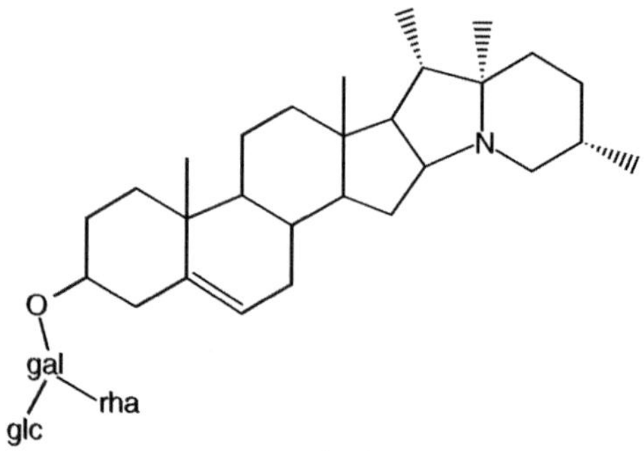

α-solanine

Toxicity

Cholestanes are potent cholinesterase inhibitors, and there is evidence of teratogenicity (birth defects) and hepatotoxicity (Jiang *et al.*, 2016). Avoidance of sprouting potatoes and unripe tomatoes is recommended, particularly for pregnant women.

Solasodine, a glycoalkaloid used in production of contraceptives, is obtained from *Solanum laciniatum* and *Solanum aviculare* (Solanaceae), native to Australia and New Zealand. These species are now cultivated in several countries commercially. Apart from anti-inflammatory actions consistent with other steroidal compounds, recent research into steroidal saponins has focused on anti-tumour and cancer preventative activities (Jiang *et al.*, 2016; Dey *et al.*, 2019).

The *Veratrum* genus (Melanthiaceae) is also rich in steroidal alkaloids. They are derived from the white false hellebore (*Veratrum album*) and the green false hellebore (*Veratrum viride*). These alkaloids have steroid skeletons, though some are highly oxygenated. The **protoveratrines** A and B are hexacyclic (six rings) and contain up to nine oxygen atoms.

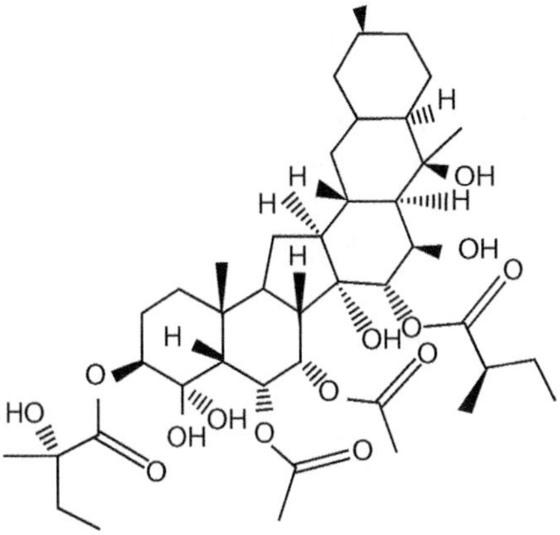

protoveratrine A

Veratrum alkaloids are cardiac depressants and have been used for treatment of severe hypertension. Their action is mediated through activation of sodium ion (Na⁺) channels and depolarization of neuronal membranes (Wink and van Wyk, 2008). Owing to their high toxicity the *Veratrum* genus is not commonly used by herbalists.

Alkaloidal amines

Amines are simple compounds derived from ammonia (NH_3) in which one or more hydrogen atoms is replaced by carbon. Replacement of one, two or three hydrogen atoms results in primary, secondary and tertiary amines, respectively. Amino acids and alkaloids are derived from amines. In alkaloidal amines, however, the only nitrogen atoms occur in the amino side group attached to a benzene ring – they are not heterocyclic and are often regarded as 'pseudo alkaloids'.

The precursors for alkaloidal amines are the aromatic amino acids – **phenylalanine, tyrosine** and **tryptophan.**

The first alkaloid formed from phenylalanine is **cathionine**, the active constituent of the drug khat, derived from fresh leaves of *Cathis edulis* (Celastraceae), a popular beverage in parts of Africa and Middle Eastern countries. Reduction and methylation of cathionine lead to **ephedrine** and **pseudoephedrine**, which are derived from the aerial parts of *Ephedra* spp. (Ephedraceae). The major sources of these alkaloids are the oriental species, including ma huang (*Ephedra sinica*), which for centuries has been a leading medicine in the *Chinese Materia Medica*. Other oriental sources are *Ephedra intermedia* and *Ephedra equisetina*, both of which are used in traditional Japanese medicine (kampo) (Yoshimura *et al.*, 2020). In the West, use of the herb, ephedra (including several North American species with low alkaloid content) is clouded in controversy with official restrictions in many countries (González-Juárez *et al.*, 2020). The two major alkaloids in *Ephedra* form the basis of several proprietary prescription medicines, as well as in illicit amphetamine drugs, including **MDMA** (ecstasy).

Ephedrine is structurally simple, its aromatic skeleton deriving from phenylalanine. The basic structure occurs in several isomeric forms, one of which (a diastereoisomer) is pseudoephedrine.

ephedrine – an alkaloidal amine　　　　cathionine

Ephedrine is a sympathomimetic or central nervous system stimulant. It is a mediator of α and β adrenergic receptors, either by direct interaction or indirectly by releasing catecholamines (i.e. noradrenaline) from synapses and/or inhibiting their reuptake (Ma *et al.*, 2007). The effects include vasoconstriction, bronchodilation, diuresis and a rise in both blood pressure and pulse. Ephedrine-based drugs are used as nasal decongestants, bronchodilators and in anaphylactic shock. In excess they cause insomnia, palpitations, tachycardia and dysuria (Odaguchi *et al.*, 2018).

In herbal medicine, *Ephedra* is valued as a reliable treatment for asthma and allergic conditions of all types, being diaphoretic, bronchodilatory and a nasal decongestant (due to constriction of blood vessels). *Ephedra* herb is also useful in oedema, headache, bronchitis, emphysema and rhinitis, as well as common colds and influenza. It is used for weight loss and energy enhancement, often at dangerously high doses, prompting the US Federal Drug Authority to ban the sale of all *Ephedra* and ephedrine-containing drugs (González-Juárez *et al.*, 2020). The herb is contraindicated for hypertension, angina pectoris, hyperthyroidism, pregnancy and where monoamine oxidase inhibitors are being used.

In Japan, ephedrine alkaloids-free *Ephedra* herb extract (EFE) is currently being promoted, as there is evidence that many of the therapeutic benefits of the herb can be marshalled without the adverse effects of the ephedrine alkaloids (Hyuga *et al.*, 2016; Yoshimura *et al.*, 2020). In a safety evaluation involving human subjects as well as mice, no significant difference was observed between the traditional alkaloid-containing extract and EFE and the adverse effects that were reported were mild for both extracts (Odaguchi *et al.*, 2018).

Other alkaloidal amines

Colchicine is derived from the autumn crocus (*Colchicum autumnale*, Colchicaceae). Its unique alkaloid structure, sometimes placed in the isoquinolenes, is derived from phenylalanine and tyrosine, the nitrogen atom occuring on a side chain. Colchicine also occurs in some plants of the Liliaceae family, including the popular garden plant the gloriosa lily (*Gloriosa superba*). Colchicine is beneficial in the treatment of gout, due to inhibition of macrophages which release lactic acid (Wink and van Wyk, 2008). It is also used as a preventative treatment for the hereditary inflammatory disease, familial Mediterranean fever (Cerquaglia *et al.*, 2005). Because of its ability to inhibit cell division, colchicine is used in plant breeding and genetic research.

Mescaline is derived from the peyote cactus also known as mescal buttons (*Lophophora williamsii*, Cactaceae). Mescaline is a phenylethylamine derivative, whose structure resembles catecholamines such as adrenaline. In fact, adrenaline and noradrenaline are also found in *L. williamsii*. An entheogen, mescaline is used as a hallucinogen by some native North Americans.

Hordenine is derived from barley (*Hordeum vulgare*, Poaceae). Hordenine is another phenylethylamine derivative, synthesized from tyrosine via tyramine. It functions in the plant as a germination inhibitor (Dewick, 2009).

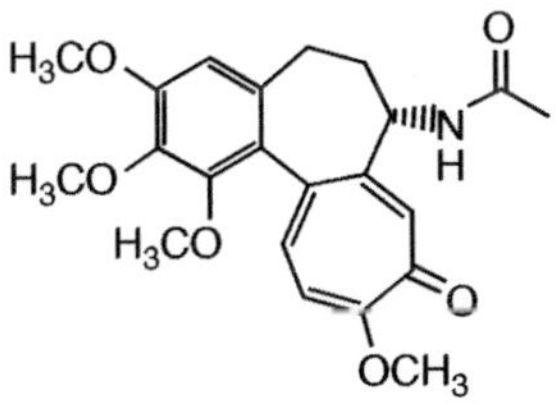

colchicine 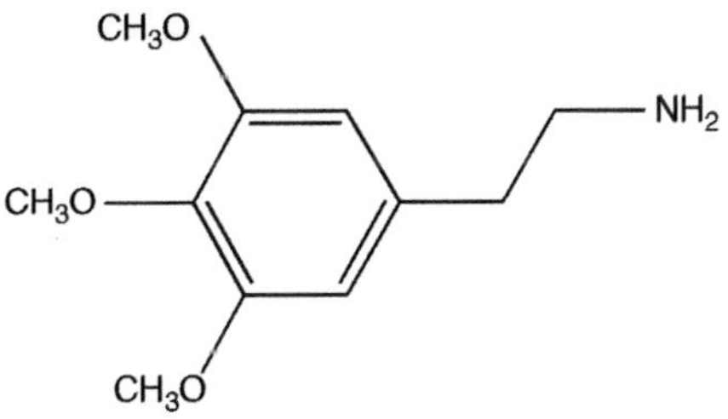   mescaline

Purine alkaloids

Purine, a dietary component mainly obtained from meat products, is a hetero-cyclic compound consisting of fused pyrimidine and imidazole rings. There are four nitrogen atoms provided by the amino acids glycine, glutamine and aspartic acid (Dewick, 2009). The purine bases adenine and guanine are components of DNA and RNA. Xanthine, an oxidized purine that occurs as a breakdown product of nucleic acid metabolism, is itself oxidized in the body to uric acid. Purine alkaloids are methylated xanthines, forming weak bases that are pharmacologically active. There are three methylxanthines and all are present in our most popular stimulant beverages – coffee and tea.

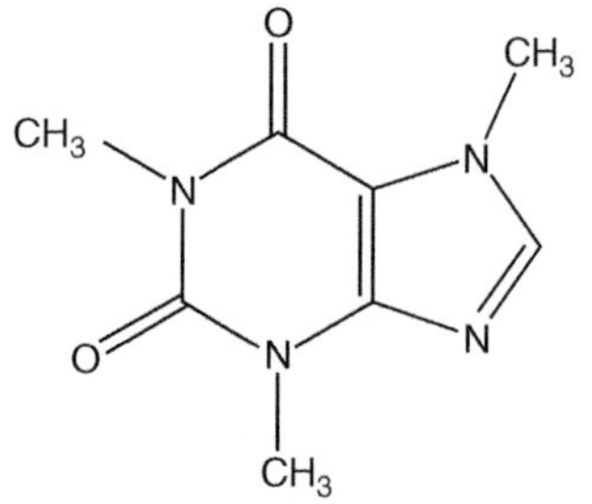

caffeine – a methylxanthine

Caffeine is found in about 100 plant species, many of which are botanically unrelated – see Table 10.2.

Coffee contains three methylxanthines: caffeine, theophylline and theobromine. Caffeine is bound to chlorogenic acid in raw coffee beans, during the roasting process the caffeine and other compounds are liberated which contributes to the aroma of coffee (Dewick, 2009).

Caffeine is a central nervous system stimulant, enhancing alertness and overcoming fatigue. It is a competitive antagonist at adenosine receptors, adenosine itself being a product of purine metabolism. Adenosine's action is to decrease neuronal activity in the brain, but when this action is blocked by caffeine, stimulating neurotransmitters (dopamine, noradrenaline, serotonin) dominate

Table 10.2. Plants noted for containing caffeine

Family	Botanical name	Common name	Caffeine (%)
Rubiaceae	*Coffea arabica*	Coffee	0.8–2.8
Theaceae	*Camellia sinensis*	Tea	1.4–4.5
Malvaceae	*Cola nitida*	Cola nut	2–4
Sapindaceae	*Paullinia cupana*	Guarana	0.23
Aquifoliaceae	*Ilex paraguariensis*	Mate	2–4.5
Aquifoliaceae	*Ilex vomitoria*	Yaupon	1–1.5

and the brain is in a state of alertness and wakefulness (Kennedy, 2014). High doses of caffeine (over 400 mg/day) are habit-forming and can lead to insomnia and tremors. In general caffeine stimulates cardiac output, blood pressure and heart rate, increases metabolism and raises blood sugar levels. Used in formulations for treating migraine, caffeine is also a mild diuretic. Research indicates that caffeine reinforces the reward effects of drugs such as nicotine and cocaine, as well as the flavour and aromas of caffeinated beverages (Kennedy, 2014).

Coffee itself has multiple health benefits, though many of these are due to the presence of polyphenols such as caffeic and cholinergic acids. Coffee consumption has been shown to increase levels of probiotic bifidobacteria in humans, thereby benefiting gut health (Jaquet *et al.*, 2009).

Structurally **theophylline** resembles caffeine; however, it lacks the methyl group in the five-carbon ring. It occurs naturally in the tea plant, but it is synthesized from caffeine for use in medicine. The effects on the central nervous system and cardiovascular system are similar to those of caffeine though milder, while the diuretic activity is more pronounced. However, the main use for theophylline is as a bronchial smooth muscle relaxant for treatment of bronchial asthma and emphysema. It forms the basis of the drug **aminophylline**, used as a diuretic and asthma medicine.

Three popular caffeine-containing beverages of South American origin are derived from the cacao plant (*Theobroma cacao*, Malvaceae), guarana (*Paullinia cupana*, Sapindaceae) and mate tea (*Ilex paraguariensis*, Aquifoliaceae). Yaupon holly (*Ilex vomitoria*) is the only caffeine-producing plant native to North America.

Theobromine is found mainly in the cacao plant. It is isomeric with theophylline; however, theobromine lacks the potent central nervous system effects of the other purine alkaloids. There is evidence linking cocoa consumption with neuroprotective benefits, though polyphenols and other constituents may be involved along with the theobromine (Cova *et al.*, 2019). Theobromine is used mainly as a diuretic and bronchial muscle relaxant.

Amino acids

Amino acids are nitrogenous compounds whose main role is the synthesis of proteins necessary for growth and maintenance of healthy tissues. They also act as precursors to alkaloids (see above). Apart from their nutritional and biosynthetic roles, amino acids are increasingly being utilized as therapeutic agents for a wide range of conditions (Marshall, 1994). While most emphasis is on the essential amino acids (primary metabolites not covered in this text), some lesser-known non-protein amino acids have also proven to be of interest. These are found mainly in the Fabaceae (legume) family.

L-canavanine is found in the seeds of many legumes, including lucerne (alfalfa) (*Medicago sativa*) and jackbean (*Canavalia ensiformis*). Lower levels of L-canavanine are found in the South African medicinal cancer bush (*Lessertia frutescens*, Fabaceae), formerly known as *Sutherlandia frutescens*. L-canavanine is an L-arginine antagonist and therefore an antimetabolite, since it replaces an essential amino acid.

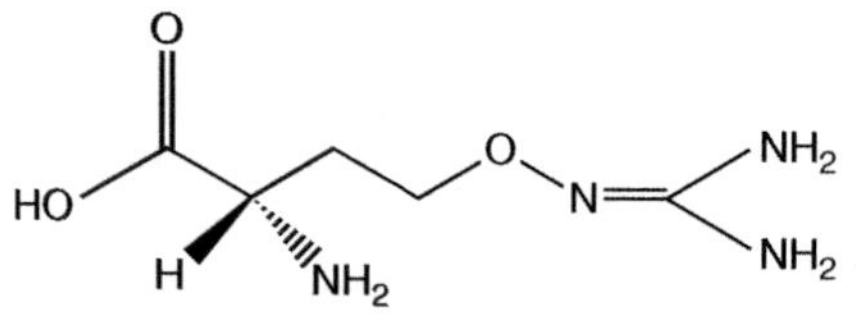

L-canavanine

L-canavanine is a documented antiviral, antibacterial, antifungal and anti-neoplastic agent. Antiviral activity has been demonstrated for both influenza and retroviruses. It is a selective inhibitor of inducible nitric oxide synthase. The antitumour mechanism involves incorporation of canavanine into the protein of cancer cells, where it replaces arginine (Rosenthal and Nkomo, 2000).

The *Lathyrus* genus (Fabaceae), well known for the sweet-pea (*Lathyrus odoratus*) and species of vetch including *Lathyrus sativus*, contain non-protein amino acids such as **oxalyminopropionic acid** and **cadaverine**, which can cause lathyrism in humans and livestock which manifests as rigidity and loss of muscle strengthen in the legs, leading towards paralysis. If the spine is affected the disorder becomes neurolathyrism (Askari, 2010).

Lectins

Lectins are small proteins consisting of two peptide chains, which bind select-ively with carbohydrates on cell surfaces, causing agglutination (clumping) or precipitation of the carbohydrates (Loris *et al.*, 1998; Wink and van Wyk, 2008). Some lectins can act as mitogens, stimulating immune cells to undergo mitotic division, proliferate and mature (Ashraf and Khan, 2003). Their large complex structures do not lend themselves to two-dimensional diagrammatic representation.

The first lectin discovered was **concanavalin A** from jack bean, and more recently it became the first lectin to have its whole protein sequence and three-dimensional structure determined (van Damme, 2014). Lectins are espe-cially common in legumes (e.g. soybeans, lentils, kidney beans). **Abrin**, a mix-ture of four lectins, is found in the seeds of the notorious garden plant rosary pea (aka crab's claw creeper) (*Abrus precatorius*, Fabaceae) – one of the dead-liest plants known (Wink and van Wyk, 2008). Equally toxic is **ricin**, also a four-lectin mixture, from seeds of the castor plant *Ricinis communis* (Euphorbiaceae). Fortunately, the seed oil (i.e. castor oil) contains no lectins. Abrin and ricin are haemagluttinins, also known as ribosome inactivating proteins (RIPs) as they inhibit protein synthesis. Symptoms following ingestion include vomiting, bloody diarrhoea, acute gastroenteritis, delirium and coma leading to death

(Wink and van Wyk, 2008). Most lectins are minimally toxic and those that occur in foods such as beans are readily destroyed by processing and cooking. While widespread within the plant kingdom, lectins are also produced by viruses and bacteria.

Medicinal species containing lectins include pokeweed (*Phytolacca americana*, Phytolaccaceae), mistletoe (*Viscum album*, Viscaceae), nettle (*Urtica dioica*, Urticaeae) and black walnut (*Juglans nigra*, Juglandaceae) (Lewis and Elvin-Lewis, 1977). Mistletoe contains lectins (**ML I–III**), viscotoxins (low-molecular-weight polypeptides), amines, polysaccharides, alkaloids, flavonoids, triterpenes, sterols, fatty acids and phenylpropanoids. ML I–III have been found to bind to erythrocytes, lymphocytes, leucocytes, macrophages, glycoproteins and plasma proteins. The lectins are RIPs, similar in action to ricin, however, ricin is about 27 times more toxic (Barnes *et al.*, 2007). The immune-stimulating effects attributed to mistletoe lectins and viscotoxins are active when injected at low concentrations (van Wyk and Wink, 2017).

Cytotoxic activity of mistletoe has been demonstrated for the glycoprotein fraction, alkaloid fraction and Iscador™ (plant juice preparation) – positive *in vitro* and *in vivo*. Human studies with Iscador™ have shown slight improvement over controls, with best results for colon cancer. In Europe it is often used as adjunct treatment to surgery and radiotherapy (Newall *et al.*, 1996).

References

Anitha, S.S. (2016) Pharmacological activity of *Vinca* alkaloids. *Journal of Pharmacognosy and Phytochemistry* 4(3), 37–44.

Ashraf, M.T. and Khan, R.H. (2003) Mitogenic lectins. *Medical Science Monitor: International Medical Journal of Experimental and Clinical Research* 9(11), RA265–RA269.

Askari, S.H.A. (2010) *Poisonous Plants of Pakistan*. Oxford University Press, Oxford, UK.

Barnes, J., Anderson, L.A. and Phillipson, A.D. (2007) *Herbal Medicines*, 3rd edn. Pharmaceutic Press, London.

Biondich, A.S. and Joslin, J.D. (2016) Coca: the history and medical significance of an ancient Andean culture. *Emergency Medicine International* 2016: 4048764. https://doi.org/10.1155/2016/4048764

Bleasel, M.D. and Peterson, G.M. (2020) Emetine, ipecac, ipecac alkaloids and analogues as potential antiviral agents for coronaviruses. *Pharmaceuticals* 13, 51. doi: 10.3390/ph13030051

Bradley, P. (ed.) (1992) *British Herbal Compendium – a Handbook of Scientific Information on Widely Used Plant Drugs*. British Herbal Medicine Association, Bournemouth, UK.

Carhart-Harris, R.L., Roseman, L., Bolstridge, M., Demetriou, L., Pannekoek, J.N., *et al.* (2017) Psilocybin for treatment-resistant depression: fMRI-measured brain mechanisms. *Scientific Reports* 7(1), 13187. https://doi.org/10.1038/s41598-017-13282-7

Cerquaglia, C., Diaco, M., Nucera, G., La Regina, M., Montalto, M. and Manna, R. (2005) Pharmacological and clinical basis of treatment of familial Mediterranean fever

(FMF) with colchicine or analogues: an update. *Current Drug Targets—Inflammation & Allergy* 4(1), 117–124. https://doi.org/10.2174/1568010053622984

Chen, L., Mulder, P.P.J., Louisse, J., Peijnenburg, A., Wesseling, S. and Rietjens, I.M.C.M. (2017) Risk assessment for pyrrolizidine alkaloids detected in (herbal) teas and plant food supplements. *Regulatory Toxicology and Pharmacology* 86, 292–306. http://dx.doi.org/10.1016/j.yrtph.2017.03.019

Conzelman, C.S. and White, D.M. (2016) *Roadmaps to Regulation: Coca, Cocaine, and Derivatives*. The Beckley Foundation, Beckley Park, Oxford, UK.

Cova, I., Leta, V., Mariani, C., Pantoni, L. and Pomati, S. (2019) Exploring cocoa properties: is theobromine a cognitive modulator? *Psychopharmacology* 236(2), 561–572. https://doi.org/10.1007/s00213-019-5172-0

Dewick, P.M. (2009) *Medicinal Natural Products: a Biosynthetic Approach*, 3rd edn. Wiley, Chichester, UK. doi: 10.1002/9780470742761

Dey, P., Kundu, A., Chakraborty, H.J., Kar, B., Choi, W.S., *et al.* (2019) Therapeutic value of steroidal alkaloids in cancer: current trends and future perspectives. *International Journal of Cancer* 145(7), 1731–1744. https://doi.org/10.1002/ijc.31965

Dietis, N., Rowbotham, D.J. and Lambert, D.G. (2011) Opioid receptor subtypes: fact or artifact? *British Journal of Anaesthesia* 107(1), 8–18. doi: 10.1093/bja/aer115

EFSA (European Food Safety Authority) Panel on Contaminants in the Food Chain (CONTAM) (2011) Scientific opinion on pyrrolizidine alkaloids in food and feed. *EFSA Journal* 9(11): 2406. https://doi.org/10.2903/j.efsa.2011.2406

Evans, W. (1990) Medicinal and poisonous plants of the Solanaceae. *British Journal of Phytotherapy* 1, 26–31.

Flores-Bocanegra, L., Raja, H.A., Graf, T.N., Augustinović, M., Wallace, E.D., *et al.* (2020) The chemistry of kratom (*Mitragyna speciosa*): updated characterization data and methods to elucidate indole and oxindole alkaloids. *Journal of Natural Products* 83(7), 2165–2177. https://doi.org/10.1021/acs.jnatprod.0c00257

Gilca, M., Gaman, L., Panait, E., Stoian, I. and Atanasiu, V. (2010) *Chelidonium majus* – an integrative review: traditional knowledge versus modern findings. *Forschende Komplementarmedizin* 17(5), 241–248. https://doi.org/10.1159/000321397

Goetz, C.G. (2011) The history of Parkinson's disease: early clinical descriptions and neurological therapies. *Cold Spring Harbor Perspectives in Medicine* 1(1): a008862. https://doi.org/10.1101/cshperspect.a008862

González-Juárez, D.E., Escobedo-Moratilla, A., Flores, J., Hidalgo-Figueroa, S., Martínez-Tagüeña, N., *et al.* (2020) A review of the *Ephedra* genus: distribution, ecology, ethnobotany, phytochemistry and pharmacological properties. *Molecules* 25(14), 3283. https://doi.org/10.3390/molecules25143283

Gurley, B.J., Fifer, E.K. and Gardner, Z. (2012) Pharmacokinetic herb–drug interactions (part 2): drug interactions involving popular botanical dietary supplements and their clinical relevance. *Planta Medica* 78, 1490–1514.

Harborne, J. and Baxter, H. (1993) *Phytochemical Dictionary*. Taylor & Francis, London.

Huang, K. (1993) *The Pharmacology of Chinese Herbs*. CRC Press, Boca Raton, Florida.

Hyuga, S., Hyuga, M., Oshima, N., Maruyama, T., Kamakura, H., *et al.* (2016) Ephedrine alkaloids-free *Ephedra* herb extract: a safer alternative to *Ephedra* with comparable analgesic, anticancer, and anti-influenza activities. *Journal of Natural Medicines* 70(3), 571–583. https://doi.org/10.1007/s11418-016-0979-z

Jacobs, E. (2020) A potential role for psilocybin in the treatment of obsessive-compulsive disorder. *Journal of Psychedelic Studies* 4(2), 77–87. doi: 10.1556/2054.2020.00128

Jaquet, M., Rochat, I., Moulin, J., Cavin, C. and Bibiloni, R. (2009) Impact of coffee consumption on the gut microbiota: a human volunteer study. *International Journal of Food Microbiology* 130(2), 117–121. https://doi.org/10.1016/j.ijfoodmicro.2009.01.011

Jiang, Q., Chen, M., Cheng, K., Yu, P., Wei, X. and Shi, Z. (2016) Therapeutic potential of steroidal alkaloids in cancer and other diseases. *Medicinal Research Reviews* 36(1), 119–143. doi: 10.1002/med.21346

Jin, Z.H., Qiu, W., Liu, H., Jiang, X.H. and Wang, L. (2018) Enhancement of oral bioavailability and immune response of ginsenoside Rh2 by co-administration with piperine. *Chinese Journal of Natural Medicines* 16(2), 143–149. https://doi.org/10.1016/S1875-5364(18)30041-4

Kapoor, L.D. (1995) *Opium Poppy: Botany, Chemistry, & Pharmacology*. Food Products Press, New York.

Kennedy, D.O. (2014) *Plants and the Human Brain*. Oxford University Press, Oxford, UK.

Khajuria, A., Thusu, N. and Zutshi, U. (2002) Piperine modulates permeability characteristics of intestine by inducing alterations in membrane dynamics: influence on brush border membrane fluidity, ultrastructure and enzyme kinetics. *Phytomedicine* 9, 224–231.

Kohnen-Johannsen, K. and Kayser, O. (2019) Tropane alkaloids: chemistry, pharmacology, biosynthesis and production. *Molecules* 24, 796. doi: 10.3390/molecules24040796

Koul, I. and Kapil, A. (1993) Evaluation of the liver protective potential of piperine, an active principle of black and long peppers. *Planta Medica* 59, 413–417.

Lee, N.R., Zheng, G., Crooks, P.A., Bardo, M.T. and Dwoskin, L.P. (2018) New scaffold for lead compounds to treat methamphetamine use disorders. *The AAPS Journal* 20(2), 29. https://doi.org/10.1208/s12248-018-0192-y

Lewis, W. and Elvin-Lewis, M. (1977) *Medical Botany*. Wiley, New York.

Loris, R., Hamelryck, T., Bouckaert, J. and Wyns, L. (1998) Legume lectin structure. *Biochimica et Biophysica Acta – Protein Structure and Molecular Enzymology* 1383(1), 9–36. https://doi.org/10.1016/s0167-4838(97)00182-9

Ma, G., Bavadekar, S.A., Davis, Y.M., Lalchandani, S.G., Nagmani, R., *et al.* (2007) Pharmacological effects of ephedrine alkaloids on human alpha(1)- and alpha(2)-adrenergic receptor subtypes. *The Journal of Pharmacology and Experimental Therapeutics* 322(1), 214–221. https://doi.org/10.1124/jpet.107.120709

Marshall, W.E. (1994) Amino acids, peptides and proteins. In: Goldberg, I. (ed.) *Functional Foods*. Chapman and Hall, New York, pp. 242–260.

Maxwell, E.A., King, I.I., Kamble, S.H., Raju, K., Berthold, E.C., *et al.* (2020) Pharmacokinetics and safety of mitragynine in beagle dogs. *Planta Medica* 86(17), 1278–1285. https://doi.org/10.1055/a-1212-5475

Mondal, A., Gandhi, A., Fimognari, C., Atanasov, A.G. and Bishayee, A. (2019) Alkaloids for cancer prevention and therapy: current progress and future perspectives. *European Journal of Pharmacology* 858: 172472. https://doi.org/10.1016/j.ejphar.2019.172472

Moreira, R., Pereira, D.M., Valentão, P. and Andrade, P.B. (2018) Pyrrolizidine alkaloids: chemistry, pharmacology, toxicology and food safety. *International Journal of Molecular Sciences* 19(6): 1668. https://doi.org/10.3390/ijms19061668

Newall, C.A., Anderson, L.A. and Phillipson, J.D. (1996) *Herbal Medicines: a Guide for Health-care Professionals*. The Pharmaceutical Press, London.

Nhiem, N.X., Quan, P.M. and Van, N.T.H. (2020) Alkaloids and their pharmacology effects from *Zanthoxylum* genus. In: Sharma, K., Mishra, K., Senapati, K.K.

and Danciu, C. (eds) *Bioactive Compounds. Intech.* Available at: https://www.intechopen.com/online-first/alkaloids-and-their-pharmacology-effects-from-zanthoxylum-genus (accessed 8 December 2020).

Och, A., Zalewski, D., Komsta, Ł., Kołodziej, P., Kocki, J. and Bogucka-Kocka, A. (2019) Cytotoxic and proapoptotic activity of sanguinarine, berberine, and extracts of *Chelidonium majus* L. and *Berberis thunbergii* DC. toward hematopoietic cancer cell lines. *Toxins* 11(9), 485. https://doi.org/10.3390/toxins11090485

Odaguchi, H., Sekine, M., Hyuga, S., Hanawa, T., Hoshi, K., *et al.* (2018) A double-blind, randomized, crossover comparative study for evaluating the clinical safety of ephedrine alkaloids-free *Ephedra* herb extract (EFE). *Evidence-Based Complementary and Alternative Medicine* 2018: 4625358. https://doi.org/10.1155/2018/4625358

Pan, Z., Qin, X.J., Liu, Y.P., Wu, T., Luo, X.D. and Xia, C. (2016) Alstoscholarisines H–J, indole alkaloids from *Alstonia scholaris*: structural evaluation and bioinspired synthesis of alstoscholarisine H. *Organic Letters* 18(4), 654–657. https://doi.org/10.1021/acs.orglett.5b03583

Pantano, F., Mannocchi, G., Marinelli, E., Gentili, S., Graziano, S., *et al.* (2017) Hepatotoxicity induced by greater celandine (*Chelidonium majus* L.): a review of the literature. *European Review for Medical and Pharmacological Sciences* 21(1 Suppl.), 46–52.

Pengelly, A. and Bennett, K. (2011) *Sanguinaria canadensis L., bloodroot. Appalachian Plant Monographs.* Appalachian Center for Ethnobotanical Studies. Available at: https://plantshoe.org/static/resources/Bloodroot/Sanguinaria%20canadensis_Extended%20Monograph.pdf (accessed 8 December 2020).

Plotkin, M.J. (2020) *The Amazon. What Everyone Needs to Know.* Oxford University Press, Oxford, UK.

Posadas, I., López-Hernández, B. and Ceña, V. (2013) Nicotinic receptors in neurodegeneration. *Current Neuropharmacology* 11(3), 298–314. https://doi.org/10.2174/1570159X11311030005

Rasoanaivo, P., Wright, C.W., Willcox, M.L. and Gilbert, B. (2011) Whole plant extracts versus single compounds for the treatment of malaria: synergy and positive interactions. *Malaria Journal* 10 (Suppl. 1), S4. https://doi.org/10.1186/1475-2875-10-S1-S4

Roberts, M.F., Strack, D. and Wink, M. (2010) Biosynthesis of alkaloids and betalains. In: Wink, M. (ed.) *Biochemistry of Plant Secondary Metabolism*, 2nd edn. Wiley Blackwell, Hoboken, New Jersey, pp. 20–91.

Roddick, J. (1991) The importance of the Solanaceae in medicine and drug therapy. In: Hawkes, J.G., Lester, R.N. and Skelding, A.D. (eds) *Solanaceae III: Taxonomy, Chemistry, Evolution.* Royal Botanic Gardens, Kew, Richmond, UK, pp. 7–23.

Rode, D. (2002) Comfrey toxicity revisited. *Trends in Pharmacological Sciences* 23 (11), 497–499.

Roeder, E., Wiedenfeld, H. and Edgar, J.A. (2015) Pyrrolizidine alkaloids in medicinal plants from North America. *Pharmazie* 70, 357–367.

Roseman, L., Nutt, D.J. and Carhart-Harris, R.L. (2018) Quality of acute psychedelic experience predicts therapeutic efficacy of psilocybin for treatment-resistant depression. *Frontiers in Pharmacology* 8, 974. https://doi.org/10.3389/fphar.2017.00974

Rosenthal, G.A. and Nkomo, P. (2000) The natural abundance of L-canavanine, an active anticancer agent in alfalfa, *Medicago sativa. Pharmaceutical Biology* 38, 1–6.

Samuelsson, G. (1992) *Drugs of Natural Origin*. Swedish Pharmaceutical Press, Stockholm.

Sanchez-Ramos, J.R. (2020) The rise and fall of tobacco as a botanical medicine. *Journal of Herbal Medicine* 22: 100374. https://doi.org/10.1016/j.hermed.2020.100374

Schmeller, T., Latz-Bruning, B. and Wink, M. (1997) Biochemical activities of berberine, palmatine and sanguinarine mediating chemical defence against microorganisms and herbivores. *Phytochemistry* 44, 257–266.

Shang, X.F., Morris-Natschke, S.L., Yang, G.Z., Liu, Y.Q., Guo, X., *et al.* (2018) Biologically active quinoline and quinazoline alkaloids part II. *Medicinal Research Reviews* 38(5), 1614–1660. https://doi.org/10.1002/med.21492

Speisky, H., Squella, J. and Nunez-Vergara, L. (1991) Activity of boldine on rat ileum. *Planta Medica* 57, 519–522.

Suhaimi, F.W., Yusoff, N.H., Hassan, R., Mansor, S.M., Navaratnam, V., *et al.* (2016) Neurobiology of kratom and its main alkaloid mitragynine. *Brain Research Bulletin* 126(Part 1), 29–40. https://doi.org/10.1016/j.brainresbull.2016.03.015

Tian, Y., Zhang, C. and Guo, M. (2017) Comparative study on alkaloids and their antiproliferative activities from three *Zanthoxylum* species. *BMC Complementary and Alternative Medicine* 17(1), 460. https://doi.org/10.1186/s12906-017-1966-y

Tillhon, M., Guamán Ortiz, L.M., Lombardi, P. and Scovassi, A.I. (2012) Berberine: new perspectives for old remedies. *Biochemical Pharmacology* 84(10), 1260–1267. https://doi.org/10.1016/j.bcp.2012.07.018

Upton, R. and Petrone, C. (2019) *Boneset aerial parts Eupatorium perfoliatum L. American Herbal Pharmacopoeia and Therapeutic Compendium monograph*. American Herbal Pharmacopoeia, Scotts Valley, California.

Van Damme, E.J. (2014) History of plant lectin research. *Methods in Molecular Biology* 1200, 3–13. https://doi.org/10.1007/978-1-4939-1292-6_1

Van Wyk, B. and Wink, M. (2017) *Medicinal Plants of the World*. CAB International, Wallingford, UK.

Wang, S. (2019) Historical review: opiate addiction and opioid receptors. *Cell Transplantation* 28(3), 233–238. doi: 10.1177/0963689718811060

Wink, M. and van Wyk, B. (2008) *Mind-altering and Poisonous Plants of the World*. Briza Publications, Pretoria, South Africa.

Wren, R.C. (1988) *Potter's New Cyclopaedia of Botanical Drugs and Preparations*. C.W. Daniel Co., Saffron Walden, UK.

Yoshimura, M., Amakura, Y., Hyuga, S., Hyuga, M., Nakamori, S., *et al.* (2020) Quality evaluation and characterization of fractions with biological activity from *Ephedra* herb extract and ephedrine alkaloids-free *Ephedra* herb extract. *Chemical and Pharmaceutical Bulletin* 68(2), 140–149. https://doi.org/10.1248/cpb.c19-00761

Zhao, D., Hamilton, J.P., Pham, G.M., Crisovan, E., Wiegert-Rininger, K., *et al.* (2017) *De novo* genome assembly of *Camptotheca acuminata*, a natural source of the anti cancer compound camptothecin. *GigaScience* 6(9): gix065. https://doi.org/10.1093/gigascience/gix065

Zhou, J., Chan, L. and Zhou, S. (2012) Trigonelline: a plant alkaloid with therapeutic potential for diabetes and central nervous system disease. *Current Medicinal Chemistry* 19(21), 3523–3531. https://doi.org/10.2174/092986712801323171

Zielińska, S., Jezierska-Domaradzka, A., Wójciak-Kosior, M., Sowa, I., Junka, A. and Matkowski, A.M. (2018) Greater celandine's ups and downs – 21 centuries of medicinal uses of *Chelidonium majus* from the viewpoint of today's pharmacology. *Frontiers in Pharmacology* 9, 299. https://doi.org/10.3389/fphar.2018.00299

Plant Lipids and Alkylamides 11

Introduction

Plant lipids are classified as primary metabolites, and therefore essential for life. Unlike secondary metabolites, lipids are universally present in plants – particularly in their seeds – varying only in their abundance and chemical composition. All lipids are composed of a hydrocarbon skeleton with one or more oxygen (O) substitutes. Plant lipids are derived from the acetate pathway via molonyl CoA, a pathway that leads to fatty acids, polyketides, polyacetylenes, phospholipids, prostaglandins and alkylamides. The more complex lipids may contain elements such as phosphorus, nitrogen or sulfur.

Fixed oils

The intervention of the enzyme, fatty acid synthase, into the acetate pathway leads to the formation of fatty acids. With the addition of a sequence of two carbon units derived from acetyl-CoA **palmitic acid**, a C_{16} fatty acid is formed, this is a fixed oil. Subsequently specific enzymes add more two carbon units and double bonds, extending the aliphatic chain, thereby producing a series of fixed oils (Table 11.1) with up to 24 carbon atoms (Quinn, 2011). Fixed oils are composed of fatty acids, hydrocarbon chains with a methyl (CH_3) group at one end, the Ω or ω (omega) end, and a carboxyl group (COOH) at the other, α (alpha – sometimes referred to as delta Δ) end. The fatty acids are esterified with glycerol to form triglycerides (three fatty acid units to one glycerol). Depending on the length of the carbon chain, fatty acids are classified as short chain fatty acids (SCFA), medium chain fatty acids (MCFA) or long chain fatty acids (LCFA).

  DOI: 10.1079/9781789243079.0011

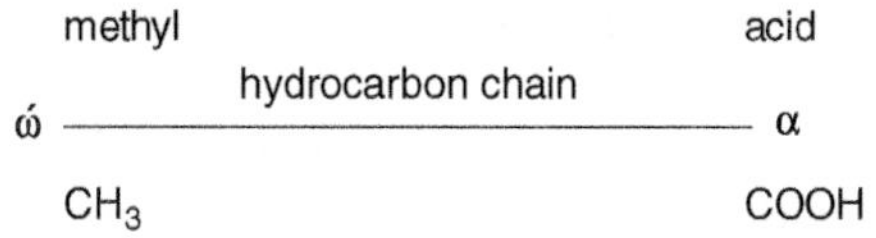

methyl   acid

ꞷ ——————— hydrocarbon chain ——————— α

CH$_3$ COOH

a simplified fatty acid structure

Table 11.1. Selective list of fixed oils

Name	Carbon chain	Double bonds	Carbon chain type	Major source
Buytric acid	4C	0	SCFA	Butterfat
Caproic acid	6C	0	SCFA	Coconut oil
Caprylic acid	8C	0	MCFA	Butterfat
Capric acid	10C	0	MCFA	Butterfat
Lauric acid	12C	0	MCFA	Coconut oil
Myristic acid	14C	0	MCFA	Coconut oil
Palmitic acid	16C	0	LCFA	Peanut oil
Palmitoleic acid	16C	1	LCFA	Palm kernel oil
Stearic acid	18C	0	LCFA	Butterfat
Oleic acid	18C	1	LCFA	Olive oil
Linoleic acid	18C	2	LCFA	Soybean oil
α-Linolenic acid	18C	3	LCFA	Rapeseed oil
γ-Linolenic acid	18C	3	LCFA	Evening primrose oil
Arachidonic acid	20C	4	LCFA	Eggs
Eicosapentaenoic acid (EPA)	20C	5	LCFA	Fish oils
Benehic acid	22C	0	LCFA	Peanut oil
Docosahexaenoic acid (DHA)	22C	6	LCFA	Fish oils
Nervonic acid	24C	1	LCFA	Honesty seed oil

LCFA, long chain fatty acid; MCFA, medium chain fatty acid; SCFA, short chain fatty acid.

Carbon bonding

Fatty acids are classified according to the number of double bonds, and which end of the carbon chain the double bonds are nearest, for example:

- saturated fatty acids – no double bonds, e.g. stearic acid;
- mono-unsaturated fatty acids – one double bond, e.g. oleic acid; and
- polyunsaturated fatty acids – two or more double bonds, e.g. linoleic acid.

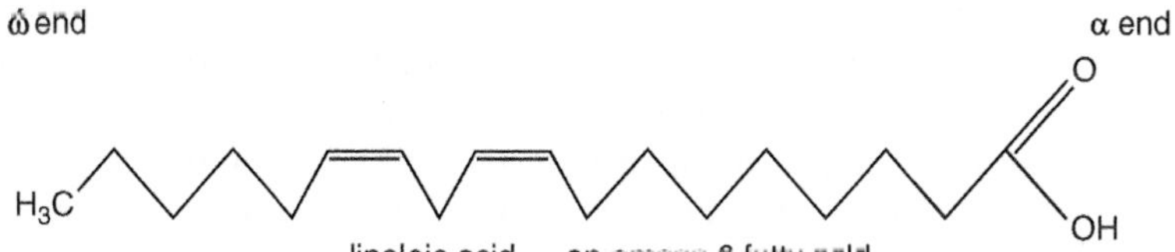

ꞷ end   α end

H$_3$C   O

        OH

linoleic acid — an omega 6 fatty acid

Structure of a polyunsaturated acid

The more double bonds a lipid contains, the greater the fluidity. Conversely saturated fatty acids without double bonds tend to be solid. Plants and animals from cold regions (e.g. fish from the deep oceans) contain high levels of poly-unsaturated acids whereas those from tropical regions (e.g. pigs, coconut and palm oils) contain more saturated fatty acids.

Nomenclature of fatty acids

Shorthand conventions are widely used to represent the various fatty acids. Carbon atoms may be counted from either end of the hydrocarbon chain; however, in this chapter they are all counted from the methyl end. **Linoleic acid** (LA) has 18 carbons with two double bonds, the first of which occurs at the sixth carbon from the ω (or Ω), or methyl end. Hence it is represented as 18:2 ω6 (or 18:2 n-6), one of the family of n-6 fatty acids. Note that in general writing the letter 'n' denotes ω, therefore, LA is an n-6 fatty acid.

Omega-3 and -6 essential fatty acids

While, in general, fatty acids are essential for human growth and survival, there are two specific compounds the body is unable to synthesize – therefore, they must be obtained through the diet. The two essential fatty acids (EFAs) are linoleic acid and **α-linolenic acid** (ALA) (18:3 ω3). EFAs are present in all the body's cells, and especially concentrated in the brain and central nervous system. EFAs are needed for regulation of immune function, vision, cell membrane stability, and production of eicosanoids such as prostaglandins (which regulate pain, temperature, blood pressure) and leukotrienes (for defence against infections).

Following ingestion in the diet, EFAs are metabolized to longer chain fatty acids. In the case of Δ-6-desaturase, LA and ALA both compete for this enzyme. The ratio between the metabolites of the two pathways influences the balance of anti- and pro-inflammatory prostaglandins. Pro-inflammatory prostaglandins include: (i) prostaglandin (PG) series 2 – derived from arachidonic acid (AA) via cyclooxygenase (COX); and (ii) thromboxane and leucotrienes derived via lipoxygenase (LOX) – see metabolic pathways shown below (Calder, 2017; Timoszuk *et al.*, 2018). These prostaglandins are further transformed to **hydroperoxyeicosatraenoic acid** (HPETE) and **hydroxyeicosatraenoic acid** (HETE), which depending on the LOX isoform, may be pro- or anti-inflammatory (Timoszuk *et al.*, 2018).

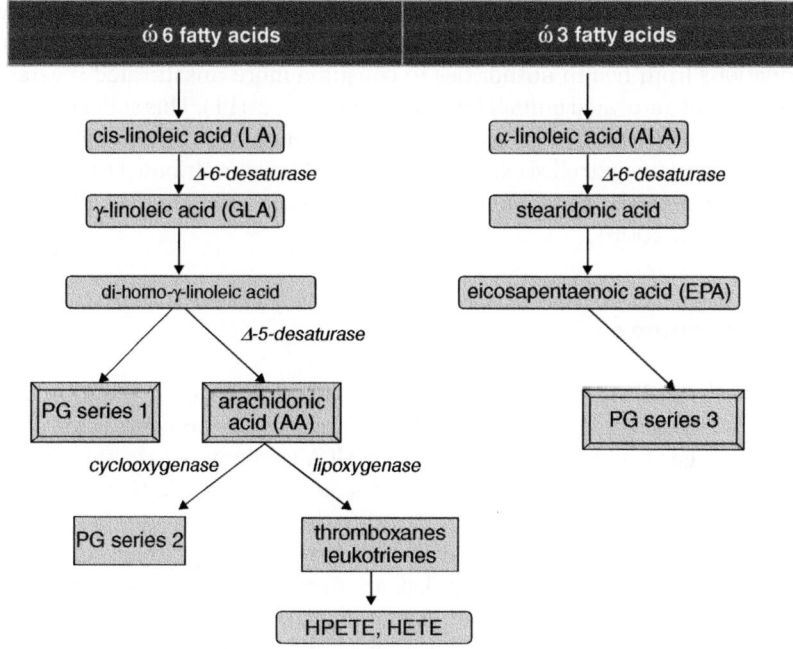

metabolic pathways of n-6 and n-3 essential fatty acids – ('pathways to inflammation') also known as the arachidonic acid pathway. Metabolizing enzymes are named in *italics*

Table 11.2 lists some of the main dietary sources of EFAs. The quality of these dietary sources depends not only on the levels of the essential fatty acids, but also on their n-3:n-6 ratios. Many vegetable oils contain moderate levels of saturated fatty acids such as palmitic acid – they, too, provide therapeutic benefits.

In modern Western diets, consumption of n-6 fatty acids in the form of LA has progressively increased at the expense of n-3s. In the USA this has been linked to an extraordinary increase in consumption of soybean oil, where a

Table 11.2. Dietary sources of essential fatty acids

Linoleic acid 18:2 omega-6	α-Linolenic acid 18:3 omega-3
Sunflower seed	Deep ocean fish (e.g. salmon, mackerel)
Safflower seed	Cod liver oil
Evening primrose	Flax seed
Borage seed	Walnuts
Hemp seed	Hemp seed
Blackcurrant seed	Blackcurrant seed
Corn	Green vegetables
Legumes (e.g. soy oil)	
Lean meat especially wild game	

more than 1000-fold increase has been recorded, largely due to the massive scale of soybean production as a favoured crop for agribusiness, and recommendations from health authorities to consume more unsaturated vegetable oils instead of saturated animal fats (Blasbalg *et al.*, 2011). This sudden switch (in evolutionary terms) to LA as a major source of calories has been described as a 'very large uncontrolled experiment that may have contributed to increased societal burdens of aggression, depression, and cardiovascular mortality' (Hibbeln *et al.*, 2006).

Evening primrose oil

The oil is derived from seed of *Oenothera biennis* (Onagraceae). Evening primrose oil is rich in the n-6 fatty acids: linoleic acid and **γ-linolenic acid** (GLA). While LA is easily obtained as a dietary oil, GLA is relatively rare in foods, but humans can derive some GLA from conversion of dietary LA. However, there are many factors that may result in blocking the enzymic process responsible for this conversion. Atopic individuals with inherent susceptibility to allergic disorders, eczema and asthma included, are usually deficient in GLA (Dewick, 2009; Timoszuk *et al.*, 2018). Therapeutically, evening primrose oil and other preparations containing GLA are used for atopic eczema, premenstrual syndrome and mastalgia, rheumatoid arthritis, endometriosis and schizophrenia (Vaddadi, 1981; Wright and Burton, 1982). GLA is also present in oil of borage seed (*Borago officinalis*) and the seeds of other plants in the Boraginaceae, as well as species in the blackcurrant family – Grossulariaceae (see Table 11.3).

While it is obvious from Table 11.3 that evening primrose oil is by no means the highest source of GLA, it is often preferred due to the corresponding high levels of LA. However, it contains virtually no ALA, whereas seeds from species of *Ribes* (Grossulariaceae) contain relatively high amounts (McKenna *et al.*, 2002). These factors must be considered when prescribing essential fatty acids for therapeutic use – often, a combination of evening primrose oil with a fish oil supplement is recommended (see below).

Table 11.3. Botanical sources of γ-linolenic acid (GLA) (from McKenna *et al.*, 2002)

Botanical name	Common name	GLA (%)
Borago officinalis	Borage	20.0–26.0
Pectocarya platycarpa	Broadfruit combseed	15
Ribes nigrum	Blackcurrant	14.0–19.0
Alkanna orientalis	Alkanet	12.4
Ribes uva-crispa	Gooseberry	10.0–12.0
Scrophularia marilandica	American figwort	9.6
Oenothera grandiflora	Large evening primrose	9.3
Oenothera biennis	Evening primrose	8.0–12.0

Eicosapentaenoic acid

Eicosapentaenoic acid (EPA) is an n-3 fatty acid metabolized in the body from ALA. EPA can also be obtained from fish oils. This long-chain (20-carbon) poly-unsaturated fatty acid inhibits pro-inflammatory prostaglandins including leukotrienes – responsible for bronchospasms that occur in asthma. Clinical studies with EPA/DHA supplements report reduced pain, stiffness, number of tender joints and use of NSAIDs (non-steroidal anti-inflammatory drugs) in patients with rheumatic arthritis (Calder, 2017).

High levels of dietary EPA reduce thrombosis by activation of a blood-clotting inhibitor, prostaglandin, and partial inactivation of the platelet ag-gregation factor. The resulting decrease in blood viscosity leads to prolonged bleeding time and thus EPA should be taken with caution by those on an antithrombotic therapy. Other beneficial effects are decreases in cholesterol, VLDL (very low density lipoprotein) and triglycerides, blood pressure, heart rate and heart rate variability, platelet aggregation, endothelial function and inflammation. Taken together, these actions indicate that EPA is generally protective against diseases of the vascular system (Innes and Calder, 2020). Despite these benefits, a recent Cochrane systematic review of n-3 fatty acids found they provide no benefit for prevention of cardiovascular diseases (Abdelhamid *et al.*, 2018).

Saw palmetto – *Serenoa repens* (Arecaceae)

Fruits obtained from *S. repens*, a dwarf palm of North American origin, are very rich in free fatty acids. These include oleic, lauric, myristic, capric, palm-itic, stearic and linoleic fatty acids as well as methyl esters of these fatty acids. Other lipophilic constituents include polyprenols, carotenoids, tocopherols and volatile oils (Marti *et al.*, 2019). Water-soluble polysaccharides with high molecular weights are also found in the seed.

The lipophilic compounds from this herb, most notably, lauric acid, inhibit the enzyme (5α-reductase) responsible for converting the male hormone, testosterone, to 5α-dihydrotestosterone, the metabolite associated with pros-tate enlargement. Clinical trials support the use of saw palmetto extracts in the treatment of benign prostatic hypertrophy (BPH) and lower urinary tract symp-toms (LUTS) while its anti-androgenic properties may also assist in acne, female hirsutism and even baldness. *S. repens* improves urine flow values, residual urine values and all other symptoms of BPH to a similar degree as standard drugs, such as finasteride. However, saw palmetto does not have the side effects of standard drugs, such as decreased libido or modification of PSA (prostate-specific antigen) values, the marker antigen for prostate cancer (Cristoni *et al.*, 2000). In a recent systematic review and meta-analysis *S. repens* was found to be as effective in reducing BPH and LUTS symptoms as the selective α_1-adrenaline receptor blocker drug Tamsulosin, and without any of the side effects related to sexual function associated with that drug (Cai *et al.*, 2020).

Investigations show that inhibition of 5α-reductase corresponds with the degree of saturation of the fatty acids, the length of the carbon chain and the absence of esterification. Esters of lauric acid showed no inhibition (Niederprum *et al.*, 1994).

Alkylamides

Amides are compounds derived from carboxylic acids and amines, involving elimination of water (similar to ester formation from acids and alcohol). Amide functional groups are quite resistant to hydrolysis, and amide linkages between amino acids and peptides are essential to the stability of proteins. **Paracetamol**, a well-known anti-inflammatory drug, is a simple amide formed from 4-hydroxy-phenylamine and acetic acid. Capsicum oleoresin contains several phenolic amides including **capsaicin**. **Anandamide**, the endogenous cannabinoid agonist, is an ethanolamide of the 20-carbon polyunsaturated arachidonic acid (Dewick, 2009).

In alkylamides, also known as N-alkylamides, amines are combined with unsaturated fatty acids by amide linkages, forming unbranched chains with one or more double and/or triple bonds. Given their fatty origins, alkylamides are insoluble in water, and partially soluble in the ethanol/water mixtures used to manufacture herbal extracts and tinctures (Bruni *et al.*, 2018). They have been reported from 25 different plant families (Veryser, 2016).

Alkylamides are responsible for the sharp, burning or tingling taste associated with herbs and spices such as prickly ash bark (*Zanthoxylum* spp., Rutaceae), black pepper (*Piper nigrum*, Piperaceae), Sichuan pepper (*Zanthoxylum* spp.) *Echinacea angustifolia*, *Echinacea purpurea* (Asteraceae) and cayenne (*Capsicum* spp., Solanaceae).

Isobutylamides

Isobutylamides are a subclass of alkylamides based on the amine group 2-methylpropyl. They first aroused the interest of researchers for their insecticidal activities, being toxic to numerous classes of insects including the ubiquitous housefly and mosquito. Upon further investigation, it was obvious that the most active insecticidal compounds were the ones that produced the most potent sialagogue (stimulating saliva flow) effects in humans (Brinker, 1992). Most isobutylamides so far investigated are derived from four plant families – Asteraceae, Rutaceae, Piperaceae and Aristolochiaceae (see Table 11.4).

Table 11.4. Medicinal plants containing isobutylamides

Asteraceae family	From other families
Echinacea angustifolia, Echinacea purpurea *Spilanthes oleracea, Spilanthes acmella* *Achillea millefolium* *Chamaemelum nobile*	*Piper nigrum, Piper longum* *Zanthoxylum clava-herculis, Zanthoxylum americanum* and other *Zanthoxylum* spp.

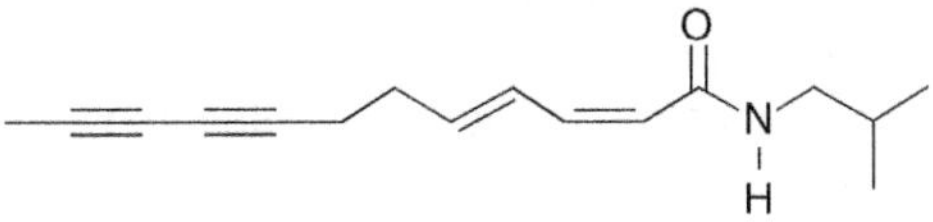

dodeca-2Z,4E-diene-8,10-diynoic acid isobutylamide
from *Echinacea angustifolia* and *E. purpurea*

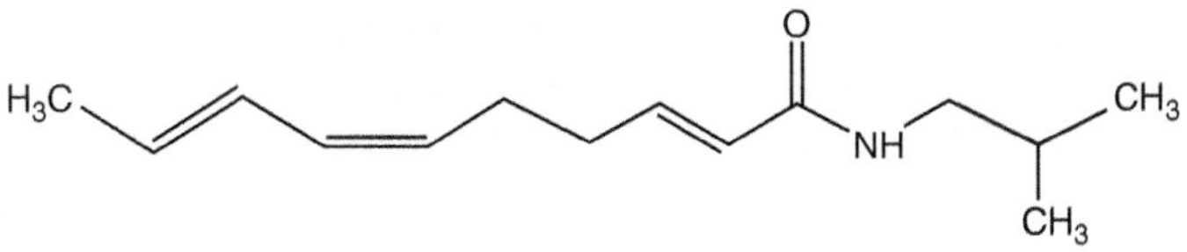

deca-2E,6Z,8E-trienoic acid isobutylamide (spilanthol) from *Spilanthes* spp.

The correct identity of the three medicinal species of *Echinacea* can be determined by assessing the differences in their chemical structures, and relative abundance of isobutylamides in the roots. The subgroup known as **(2E,4E,8Z,10E/Z)-tetraeonic acid** isobutylamides contains the key constituents. They are C_{11}–C_{16} straight-chain fatty acids with olefinic (double) and/or acetylenic (triple) bonds. Along with the phenolic compound **cynarin**, this isobutylamide type was recently confirmed as a biomarker compound for *E. angustifolia* roots (Aiello *et al.*, 2020).

According to investigations by Bauer and Wagner (1991), the relative content of these isobutylamides is as follows:

E. angustifolia	0.009–0.151%
E. purpurea	0.004–0.039%
E. pallida	negligible

Echinacea pallida lacks the distinctive sharp taste of the other two species, hence it is considered by many to be inferior. It does, however, contain its own chemical fingerprint in the form of another group of lipid-soluble constituents – **polyacetylenes**.

Pharmacology of isobutylamides

Alkylamides can be readily detected by organoleptic means, by placing a very small quantity on the tongue. The initial sharp sensation and saliva production is followed by a local anaesthetic or numbing effect. Not surprisingly many are associated with management of toothache – *Spilanthes* spp. and some *Zanthoxylum* spp. are both known in their respective regions as 'toothache plant'.

The most significant actions of alkylamides are sialagogue, analgesic, anti-inflammatory, counter-irritant, antimicrobial, insecticidal, anti-inflammatory, immune-modulating properties, vermifuge, digestive and circulatory stimulation. Alkylamide-containing herbs have been traditionally used to treat toothaches, skin and digestive diseases, viral infections and in topical cosmetics for their wrinkle smoothing and anti-ageing properties (Veryser, 2016). Antiviral actions for *E. purpurea* are widely reported, however, this action is thought to be due to a combination of constituents, most likely including alkylamides, polyphenols and possibly polysaccharides (Hudson and Vimalanathan, 2011).

Immunomodulation

Most of the research in the field of immunomodulation is focused on the broad-spectrum activities of *E. purpurea*. Along with the antiviral actions referred to previously, anti-inflammatory actions are also well documented. In one study, *E. purpurea* alkylamides were found to inhibit both isoforms of cyclooxygenase; however, inhibition of COX-1 was greater than for COX-2 (Clifford *et al.*, 2002). In an *in vitro* study involving alkylamides from *E. angustifolia* and several *Achillea* spp., 20 compounds were found to be relatively potent inhibitors of cyclooxygenase, while a few also inhibited 5-lipoxygenase (Müller-Jakic *et al.*, 1994). These two enzymes are involved in the metabolism of AA, the major pathway to inflammation (see above). Considering the structural similarities to AA, the authors proposed the alkylamides act as analogues, competitively inhibiting the enzymes, while subsequent research confirmed these compounds interact with cannabinoid receptors type-2, which are associated with immunomodulatory activity (Balan *et al.*, 2016).

In one study *E. purpurea* counteracted the reduction in lymphocytes and monocytes following exposure to radiation, by increasing production of γ-interferon and stimulation of T lymphocytes. These, in turn, stimulated macrophages to release cytokines, further enhancing the profileration of healthy helper T cells (Mishima *et al.*, 2004).

Investigations by Balan and co-workers compared different *E. purpurea* preparations for their immunomodulatory effects in mice, and found that different preparations varied between stimulating, depressing and having no effect on antibody production (Balan *et al.*, 2016). Further variations in effects for *Echinacea* have been observed with different species, different plant parts, choice of solvent, fresh versus dried plants, length and temperature of storage and other factors (Catanzaro *et al.*, 2018; Perry and van Klink, 2000).

Capsaicinoids

These compounds are responsible for the hotness in chilli peppers, the degree of pungency being related to the length of the acid side chain. Total capsaicinoid content is around 1% of the dried fruit, the majority of which is usually **capsaicin**. These compounds are so pungent they can be detected by human taste buds in solutions of ten parts per million (Zachariah and Gobinath, 2008). Structurally, capsaicinoids are vanillyl-acyl amide analogues, distinguished by the presence of a vanillin amine moiety (Tucker and Debaggio, 2000).

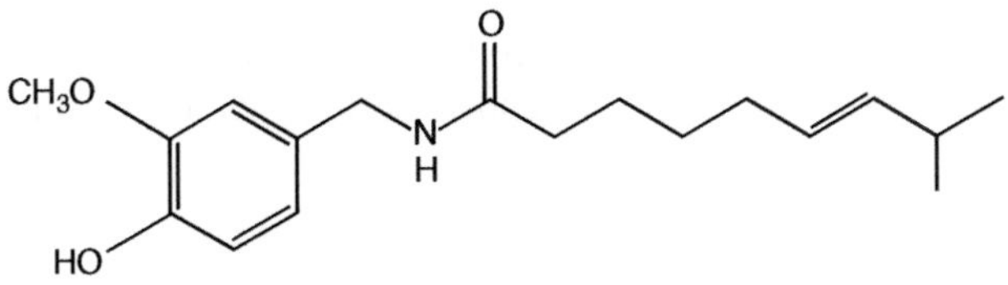

capsaicin – an aromatic amide

Capsaicinoids stimulate TRPV1 receptors (known as vanilloid receptors) on cutaneous sensory neurones, resulting in a massive release of neuropeptides including so-called 'substance P' molecules responsible for pain transmission to the brain and modulation of local inflammatory responses. Topical applications of medications containing capsaicinoids deplete the neuropeptides, therefore preventing transmission of pain signals to the brain (Warber, 1999). Hence the successful use of capsicum-based preparations for treatment of neuralgias, shingles, diabetic neuropathy and joint inflammation, such as the 8% capsaicin patch that is approved by health regulators in Europe and the USA (Basith *et al.*, 2016).

In many ways, capsaicin is the perfect subject to end this book with, being a molecule with links to so many categories of phytochemicals. A closer look at the structural diagram above shows the aromatic ring on the left (the 'head'), with hydroxyl groups indicating a phenolic structure derived from the aromatic amino acid phenylalanine. In the middle (the 'neck') we see the amide linkage which is derived from the amino acid valine, while on the right (the 'tail') an aliphatic chain derives from the acetate pathway via malonyl-CoA. The compound is also an alkaloidal amine or pseudo-alkaloid, due to the nitrogen (N) atom occurring outside of the carbon ring, and within the plant capsaicin is part of an oleoresin mixture. In addition, *Capsicum* spp. contain analogues (**capsinoids**) of capsaicin that lack nitrogen, and whose structures are similar to prenylpropanoids. These compounds lack pungency but are still pharmacologically active (Basith *et al.*, 2016).

More so than for other alkylamides, capsaicinoids are powerful skin and mucous membrane irritants, they are the key ingredients of pepper spray, the chemical riot control agent (Zachariah and Gobinath, 2008). Very high doses may cause hyperthermia and anaphylactic shock (Wink and van Wyk, 2008). Despite this the compounds are generally safe when used in low to moderate doses, the initial irritation upon contact or ingestion is short lived. Cayenne as medicine was introduced to the Western world by the great herbalist Samuel Thomson in the early 1800s, despite knowing nothing about its phytochemistry. Let me finish with a quote from Thompson's *New Guide to Health*:

It is no doubt the most powerful stimulant known; its power is entirely congenial to nature, being powerful in raising and maintaining that heat upon which life depends. It is extremely pungent, and when taken, sets the mouth as if it were on fire; this lasts however, but for a few minutes, and I consider it essentially a benefit, for its effects on the glands cause the saliva to flow freely, and leaves the mouth clean and moist.

(Thomson, 1835)

Interestingly, Thomson's second and third choices of herbal circulatory stimulants were ginger and black pepper, both of which contain phenolic alkylamides.

References

Abdelhamid, A.S., Brown, T.J., Brainard, J.S., Biswas, P., Thorpe, G.C., *et al.* (2018) Omega-3 fatty acids for the primary and secondary prevention of cardiovascular disease. *The Cochrane Database of Systematic Reviews* 7(7): CD003177. https://doi.org/10.1002/14651858.CD003177.pub3

Aiello, N., Marengo, A., Scartezzini, F., Fusani, P., Sgorbini, B., *et al.* (2020) Evaluation of the farming potential of *Echinacea angustifolia* DC. accessions grown in Italy by root-marker compound content and morphological trait analyses. *Plants* 9(7), 873. https://doi.org/10.3390/plants9070873

Bałan, B.J., Sokolnicka, I., Skopińska-Różewska, E. and Skopiński, P. (2016) The modulatory influence of some *Echinacea*-based remedies on antibody production and cellular immunity in mice. *Central European Journal of Immunology* 41(1), 12–18. https://doi.org/10.5114/ceji.2016.58813

Basith, S., Cui, M., Hong, S. and Choi, S. (2016) Harnessing the therapeutic potential of capsaicin and its analogues in pain and other diseases. *Molecules* 21(8), 966. https://doi.org/10.3390/molecules21080966

Bauer, R. and Wagner, H. (1991) *Echinacea* species as potential immunostimulatory drugs. In: Wagner, H. and Farnsworth, N.R. (eds) *Economic and Medicinal Plant Research*, vol. 5. Academic Press, New York, pp. 253–321.

Blasbalg, T.L., Hibbeln, J.R., Ramsden, C.E., Majchrzak, S.F. and Rawlings, R.R. (2011) Changes in consumption of omega-3 and omega-6 fatty acids in the United States during the 20th century. *The American Journal of Clinical Nutrition* 93(5), 950–962. https://doi.org/10.3945/ajcn.110.006643

Brinker, F.J. (1992) The insecticidal and therapeutic activity of natural isobutylamides. *British Journal of Phytotherapy* 2(4), 160–170.

Bruni, R., Brighenti, V., Caesar, L.K., Bertelli, D., Cech, N.B. and Pellati, F. (2018) Analytical methods for the study of bioactive compounds from medicinally used *Echinacea* species. *Journal of Pharmaceutical and Biomedical Analysis* 160, 443–477. https://doi.org/10.1016/j.jpba.2018.07.044

Cai, T., Cui, Y., Yu, S., Li, Q., Zhou, Z. and Gao, Z. (2020) Comparison of *Serenoa repens* with tamsulosin in the treatment of benign prostatic hyperplasia: a systematic review and meta-analysis. *American Journal of Men's Health* 14(2). https://doi.org/10.1177/1557988320905407

Calder, P.C. (2017) Omega-3 fatty acids and inflammatory processes: from molecules to man. *Biochemical Society Transactions* 45(5), 1105–1115. https://doi.org/10.1042/BST20160474

Catanzaro, M., Corsini, E., Rosini, M., Racchi, M. and Lanni, C. (2018) Immunomodulators inspired by nature: a review on curcumin and *Echinacea*. *Molecules* 23(11), 2778. https://doi.org/10.3390/molecules23112778

Clifford, L.J., Nair, M.G., Rana, J. and Dewitt, D.L. (2002) Bioactivity of alkamides isolated from *Echinacea purpurea*. *Phytomedicine* 9, 249–253.

Cristoni, A., Di Pierro, F. and Bombardelli, E. (2000) Botanical derivatives for the prostate. *Fitoterapia* 71, S21–S28.

Dewick, P.M. (2009) *Medicinal Natural Products: a Biosynthetic Approach*, 3rd edn. Wiley, Chichester, UK. doi: 10.1002/9780470742761

Hibbeln, J.R., Nieminen, L.R., Blasbalg, T.L., Riggs, J.A. and Lands, W.E. (2006) Healthy intakes of n-3 and n-6 fatty acids: estimations considering worldwide diversity. *The American Journal of Clinical Nutrition* 83(6), 1483S–1493S. https://doi.org/10.1093/ajcn/83.6.1483S

Hudson, J. and Vimalanathan, S. (2011) *Echinacea* – a source of potent antivirals for respiratory virus infections. *Pharmaceuticals* 4(7), 1019–1031. https://doi.org/10.3390/ph4071019

Innes, J.K. and Calder, P.C. (2020) Marine omega-3 (n-3) fatty acids for cardiovascular health: an update for 2020. *International Journal of Molecular Sciences* 21(4), 1362. https://doi.org/10.3390/ijms21041362

Marti, G., Joulia, P., Amiel, A., Fabre, B., David, B., *et al.* (2019) Comparison of the phytochemical composition of *Serenoa repens* extracts by a multiplexed metabolomic approach. *Molecules* 24(12), 2208. https://doi.org/10.3390/molecules24122208

McKenna, D.J., Jones, K. and Hughes, K. (2002) *Botanical Medicines*, 2nd edn. The Haworth Press, New York.

Mishima, S., Saito, K., Maruyama, H., Inoue, M., Yamashita, T., *et al.* (2004) Antioxidant and immuno-enhancing effects of *Echinacea purpurea*. *Biological and Pharmaceutical Bulletin* 27(7), 1004–1009. https://doi.org/10.1248/bpb.27.1004

Müller-Jakic, B., Breu, W., Probstle, A., Redl, K., Greger, H. and Bauer, R. (1994) *In vitro* inhibition of cyclooxygenase and 5-lipoxygenase by alkamides from *Echinacea* and *Achillea* species. *Planta Medica* 60, 37–40.

Niederprum, H., Schweikert, H. and Zanker, K. (1994) Testosterone 5α-reductase inhibition by free fatty acids from Sabal serrulata fruits. *Phytomedicine* 1, 127–133.

Perry, N.B. and van Klink, J.W. (2000) Alkamide levels in *Echinacea purpurea*: effects of processing, drying and storage. *Planta Medica* 66, 54–56.

Quinn, P.J. (2011) Membranes as targets of antimicrobial lipids. In: Halldor, T. (ed.) *Lipids and Essential Oils as Antimicrobial Agents*. Wiley, Hoboken, New Jersey, pp. 1–24. doi: 10.1002/9780470976623.ch1

Thomson, S. (1835) *New Guide to Health or Botanic Family Physician*. Boston Investigator, Boston, Massachusetts.

Timoszuk, M., Bielawska, K. and Skrzydlewska, E. (2018) Evening primrose (*Oenothera biennis*) biological activity dependent on chemical composition. *Antioxidants* 7(8), 108. https://doi.org/10.3390/antiox7080108

Tucker, A.O. and Debaggio, T. (2000) *The Big Book of Herbs*. Interweave Press, Fort Collins, Colorado.

Vaddadi, K.S. (1981) *Essential Fatty Acids in the Treatment of Schizophrenia*. World Congress of Biological Psychiatry, Stockholm.

Veryser, L. (2016) Analytical, pharmacokinetic and regulatory characterisation of selected plant *N*-alkylamides. PhD thesis, Ghent University, Ghent, Belgium. Available at: https://biblio.ugent.be/publication/8102683 (accessed 16 September 2020).

Warber, S. (1999) Modes of action at target sites. In: Kaufman, P.B., Cseke, L.J., Warber, S., Duke, J.A. and Brielmann, H.L. (eds) *Natural Products from Plants*. CRC Press, Boca Raton, Florida, pp. 415–440.

Wink, M. and van Wyk, B. (2008) *Mind-altering and Poisonous Plants of the World*. Briza Publications, Pretoria, South Africa.

Wright, S. and Burton, J.L. (1982) Oral evening primrose seed oil improves atopic eczema. *The Lancet* 2, 1120–1122.

Zachariah, T.J. and Gobinath, P. (2008) Paprika and chili. In: Parthasarathy, V.A., Chempakam, B. and Zachariah, T.J. (eds) *Chemistry of Spices*, pp. 260–286. CAB International, Wallingford, UK, pp. 260–286.

Index

Page numbers in **bold** type refer to tables.

Printed and bound by CPI Group (UK) Ltd, Croydon, CR0 4YY

06/08/2025

14714986-0001